Coordinate Geometry and Differential Calculus

A Self-Teaching Book for High Schools and Colleges

Kingsley Augustine

Table of Contents

CHAPTER 1
THE STRAIGHT LINE: CHARACTERISTICS AND EQUATION

The Coordinates of a Point

A point in a plane is specified by its distance from the x-axis and the y-axis. For example, a point given by $(-2, 5)$ is at a perpendicular distance of 2 units from the positive y-axis, and a perpendicular distance of 5 units from the negative x-axis. The value -2, is called the x-coordinate or abscissa, while 5 is called the y-coordinate of ordinate. Generally, a point in a plane is represented by (x, y).

Gradient of a Straight Line

The gradient of a straight line is the rate of change of y compared with x. Mathematically, the gradient of a straight line is represented by m, which is given by:

$$m = \frac{\text{Vertical change}}{\text{Horizontal change}}$$

$$= \frac{\text{Change in y}}{\text{Change in } x}$$

Hence, if (x_1, y_1) and (x_2, y_2) are two points on a line, then the gradient or slope of the line is given by:

$$m = \frac{y_2 - y_1}{x_2 - x_1}$$

Equation of a Line

If (x_1, y_1) is a point on a straight line of slope, m, then the equation of the line can be obtained as follows:

$$m = \frac{y - y_1}{x - x_1}$$

If (x_1, y_1) and (x_2, y_2) are two points on a line, then the equation of the line is given by:

$$\frac{y - y_1}{x - x_1} = \frac{y_2 - y_1}{x_2 - x_1}$$

There are different forms of the equation of a line. The gradient intercept form is expressed as:

$$y = mx + c$$

where m in the gradient, while c is the intercept on the y-axis. The general form of the equation of a straight line is expressed as:

$$ax + by + c = 0$$

Gradient of Parallel Lines

If two lines of slope m_1 and m_2 are parallel to each other, then their slopes are equal. This means that $m_1 = m_2$.

Gradient of Perpendicular Lines

Two lines are perpendicular to each other if the product of their slope/gradient is −1. This means that:

$$m_1 m_2 = -1 \quad \text{or} \quad m_1 = -\frac{1}{m_2} \quad \text{or} \quad m_2 = -\frac{1}{m_1}$$

Intercept on the x and y axes

If the equation of a line is $ax + by + c = 0$, then the x-axis intercept is obtained by substituting $y = 0$ into the equation above. This will give:

$$x\text{-intercept} = -\frac{c}{a} \quad \text{or} \quad \left(-\frac{c}{a}, 0\right)$$

Similarly, the y-axis intercept is obtained by substituting $x = 0$ into the equation above. This will give:

$$y\text{-intercept} = -\frac{c}{b} \quad \text{or} \quad \left(0, -\frac{c}{b}\right)$$

Point of Intersection of Two Lines

If two lines intersect, then the point of intersection of the line is obtained by solving the two equations of the lines simultaneously in order to obtain the values of x and y.

Distance between Two Points

If two points in an x-y plane are given by the coordinates $A(x_1, y_1)$ and $B(x_2, y_2)$, then the distance between them is given by:

$$AB = \sqrt{(x_2 - x_1)^2 + (y_2 - y_1)^2}$$

The Mid-point of a Line

If two points (x_1, y_1) and (x_2, y_2), form a straight line, then coordinates of the mid-point of the line is given by:

$$\left(\frac{x_1 + x_2}{2}, \frac{y_1 + y_2}{2}\right)$$

Division of a Line in a Given Ratio

If a point A divides a line PQ in the ratio p : q, with point P as (x_1, y_1) and point Q as (x_2, y_2), then the x-coordinate of point A is given by:

$$x = \frac{px_2 + qx_1}{p + q}$$

while the y-coordinate of point A is given by:

$$y = \frac{py_2 + qy_1}{p + q}$$

When the values of the ratio p and q are positive, then the line division is an internal division. When either p or q is negative, then the line division is an external division.

The Angle of Slope of a Line

The angle of slope which is the angle that a line makes with the positive x-axis is given by:

$$\tan\theta = m$$
$$\theta = \tan^{-1} m$$

where m is the gradient of the line.

When m is positive, θ is acute, when m is negative, θ is obtuse. A line which is parallel to the x-axis has a slope/gradient of zero, while a line which is parallel to the y-axis has a slope which is undefined.

Perpendicular Distance between a Point and a Line

The perpendicular distance between the point (x_1, y_1) and the line $ax + by + c = 0$, is given by:

$$D = \frac{|ax_1 + by_1 + c|}{\sqrt{a^2 + b^2}}$$

Angle between Two Lines

If a line AB has a gradient, m_1, and a line CD has a gradient, m_2, then the angle between the two lines AB and CD is given by:

$$\tan\theta = \frac{m_2 - m_1}{1 + m_1 m_2}$$

If $\frac{m_2 - m_1}{1 + m_1 m_2}$ is negative, then the angle between the two lines is obtuse. If it is positive, then the angle between the lines is acute. However, we can always determine the acute angle between any two lines by using the formula:

$$\tan\theta = \left|\frac{m_2 - m_1}{1 + m_1 m_2}\right|$$

Note that subtracting the acute angle from 180 gives the obtuse angle between the lines.

Examples

1. Find the gradients of the following lines passing through the points:
(a) A(2, 5) and B(4, 8)
(b) (−2, 3), (−5, −1)
(c) $(\frac{1}{2}, -\frac{2}{5})$ and $(-\frac{4}{5}, -\frac{3}{4})$
(d) (−0.25, −2.1) and (0.35, 1.5)
Solutions

(a) Gradient or slope $= \dfrac{\text{Change in } y}{\text{Change in } x} = \dfrac{y_2 - y_1}{x_2 - x_1}$

where A(2, 5) represents $(x_1\ y_1)$ and B(4, 8) represents (x_2, y_2)

Therefore,   Gradient $= \dfrac{y_2 - y_1}{x_2 - x_1}$

$$= \dfrac{8-5}{4-2}$$

$$= \dfrac{3}{2} = 1.5$$

(b) (−2, 3) represents $(x_1 \; y_1)$ and (−5, −1) represents (x_2, y_2)

Gradient $= \dfrac{y_2 - y_1}{x_2 - x_1}$

$$= \dfrac{-1-3}{-5-(-2)}$$

$$= \dfrac{-4}{-5+2}$$

$$= \dfrac{-4}{-3}$$

$$= \dfrac{4}{3} = 1\dfrac{1}{3}$$

(c) $\left(\dfrac{1}{2}, \; -\dfrac{2}{5}\right)$ represents $(x_1 \; y_1)$ and $\left(-\dfrac{4}{5}, \; -\dfrac{3}{4}\right)$ represents (x_2, y_2)

Gradient $= \dfrac{y_2 - y_1}{x_2 - x_1}$

$$= \dfrac{-\dfrac{3}{4} - \left(-\dfrac{2}{5}\right)}{-\dfrac{4}{5} - \dfrac{1}{2}}$$

$$= \dfrac{-\dfrac{3}{4} + \dfrac{2}{5}}{\dfrac{-8-5}{10}}$$

$$= \dfrac{\dfrac{-15+8}{20}}{\dfrac{-13}{10}}$$

$$= \dfrac{\dfrac{-7}{20}}{\dfrac{-13}{10}}$$

$$= \dfrac{-7}{20} \div \dfrac{-13}{10}$$

$$= \frac{7}{20} \times \frac{10}{13} \qquad \text{(Note that the negative sign has cancelled out)}$$

$$= \frac{7}{2} \times \frac{1}{13} \qquad \text{(After equal division by 10, since 20 divided by 10 is 2)}$$

$$= \frac{7}{26}$$

(d) $(-0.25, -2.1)$ represents $(x_1\ y_1)$ and $(0.35, 1.5)$ represents (x_2, y_2)

$$\text{Gradient} = \frac{y_2 - y_1}{x_2 - x_1}$$

$$= \frac{1.5 - (-2.1)}{0.35 - (-0.25)}$$

$$= \frac{1.5 + 2.1}{0.35 + 0.25}$$

$$= \frac{3.6}{0.6}$$

$$= 6$$

2. Two points on a line are $(-3, e)$ and $(4, -1)$. If the gradient of the line is 2, find the value of e.

Solution

$$m = \frac{y_2 - y_1}{x_2 - x_1}$$

$$2 = \frac{-1 - e}{4 - -3}$$

$$2 = \frac{-1 - e}{7}$$

$$2 \times 7 = -1 - e$$

$$e = -1 - 14$$

$$e = -15$$

3. Find the distance between the following points:

(a) X(2, 5) and Y(5, 1)

(b) A(−2, 4) and B(3, −1)

(c) (−3, −5) and (−4, −7)

(d) (−2, 3) and (5, −1)

Solution

(a) $XY = \sqrt{(x_2 - x_1)^2 + (y_2 - y_1)^2}$

where (2, 5) is (x_1, y_1) and (5, 1) is (x_2, y_2)

$$XY = \sqrt{(5-2)^2 + (1-5)^2}$$
$$= \sqrt{(3)^2 + (-4)^2}$$
$$= \sqrt{9+16}$$
$$= \sqrt{25}$$
XY = 5 units

(b) $AB = \sqrt{(x_2 - x_1)^2 + (y_2 - y_1)^2}$
where (−2, 4) is (x_1, y_1) and (3, −1) is (x_2, y_2)
$$AB = \sqrt{(3 - -2)^2 + (-1 - 4)^2}$$
$$= \sqrt{(5)^2 + (-5)^2}$$
$$= \sqrt{25+25}$$
$$= \sqrt{50}$$
$$= \sqrt{25} \times \sqrt{2}$$
$AB = 5\sqrt{2}$ units

(c) Let the distance between the two points be D. Hence:
$$D = \sqrt{(-4 - -3)^2 + (-7 - -5)^2}$$
$$= \sqrt{(-4+3)^2 + (-7+5)^2}$$
$$= \sqrt{(-1)^2 + (-2)^2}$$
$$= \sqrt{1+4}$$
$D = \sqrt{5}$ units
Or   D = 2.24 units

(d) $D = \sqrt{(5 - -2)^2 + (-1 - 3)^2}$
$$= \sqrt{(5+2)^2 + (-4)^2}$$
$$= \sqrt{7^2 + 16}$$
$$= \sqrt{49+16}$$
$$= \sqrt{65}$$
D = 8.1 units

4. Find the mid-points of the lines that join the following points:
(a) X(3, 4) and Y(1, 6)
(b) A(−4, 3) and B(−2, −5)
(c) (2, 3) and (5, −1)
(d) (−4, −7) and (−3, −2)
Solution

(a) The mid-point of XY = $\left(\dfrac{x_1+x_2}{2}, \dfrac{y_1+y_2}{2}\right)$

where (3. 4) is (x_1, y_1) and (1, 6) is (x_2, y_2)

$= \left(\dfrac{3+1}{2}, \dfrac{4+6}{2}\right)$

$= \left(\dfrac{4}{2}, \dfrac{10}{2}\right)$

$= (2, 5)$

(b) The mid-point of AB = $\left(\dfrac{x_1+x_2}{2}, \dfrac{y_1+y_2}{2}\right)$

$= \left(\dfrac{-4+-2}{2}, \dfrac{3+(-5)}{2}\right)$

$= \left(\dfrac{-4-2}{2}, \dfrac{3-5}{2}\right)$

$= \left(\dfrac{-6}{2}, \dfrac{-2}{2}\right)$

$= (-3, -1)$

(c) Let the mid-point of the line joining the two points be M. Hence:

$M = \left(\dfrac{2+5}{2}, \dfrac{3-1}{2}\right)$

$= \left(\dfrac{7}{2}, \dfrac{2}{2}\right)$

$= \left(\dfrac{7}{2}, 1\right)$ or   (3.5, 1)

(d) $M = \left(\dfrac{-4-3}{2}, \dfrac{-7-2}{2}\right)$

$= \left(\dfrac{-7}{2}, \dfrac{-9}{2}\right)$ or (−3.5, −4.5)

5. Find the coordinates of the point which divide the lines joining the following points in the stated ratios.

(a) A(3.4) and B(6, 1) in the ratio 1:2
(b) X(−2.3) and Y(−4, −1) in the ratio −2:5
(c) (5, 2) and (−3, 1) in the ratio 1:3 externally
(d) (−3, −1) and (−2, 5) in the ratio 2:3 internally
(e) (4, 2) and (2, −6) in the ratio 3:5 externally

Solutions

(a) This is an internal division since the values of the ratio are both positive. Let R be the point which divide the line AB in the ratio 1:2. Therefore, the x coordinate of point R is given by:

$$x = \frac{px_2 + qx_1}{p + q}$$

where 1:2 is p:q and (3, 4) is (x_1, y_1), while (6, 1) is (x_2, y_2). Hence:

$$x = \frac{(1 \times 6) + (2 \times 3)}{1 + 2}$$

$$= \frac{6 + 6}{3}$$

$$= \frac{12}{3}$$

$$= 4$$

The y coordinate of the point R is given by:

$$y = \frac{py_2 + qy_1}{p + q}$$

$$= \frac{(1 \times 1) + (2 \times 4)}{1 + 2}$$

$$= \frac{1 + 8}{3}$$

$$= \frac{9}{3}$$

$$= 3$$

Therefore the coordinates of point R are (4, 3)

(b) This is an external division since one of the values of the ratio is negative. Hence:

$$x = \frac{px_2 + qx_1}{p + q}$$

$$= \frac{(-2 \times -4) + (5 \times -2)}{-2 + 5}$$

$$= \frac{8 - 10}{3}$$

$$= \frac{-2}{3}$$

The y coordinate is given by:

$$y = \frac{py_2 + qy_1}{p + q}$$

$$= \frac{(-2x - 1) + (5 \times 3)}{-2 + 5}$$

$$= \frac{2 + 15}{3}$$

$$= \frac{17}{3}$$

Therefore the point is $(-\frac{2}{3}, \frac{17}{3})$

(c) This is an external division. Hence, one of the values of the ratio has to be negative. So, let us take the ratio to be −1:3 (Take note of the new negative sign). Hence:

$$x = \frac{px_2 + qx_1}{p + q}$$
$$= \frac{(-1 \times -3) + (3 \times 5)}{-1 + 3}$$
$$= \frac{3 + 15}{2}$$
$$= \frac{18}{2}$$
$$= 9$$

$$y = \frac{py_2 + qy_1}{p + q}$$
$$= \frac{(-1 \times 1) + (3 \times 2)}{-1 + 3}$$
$$= \frac{-1 + 6}{2}$$
$$= \frac{5}{2}$$

Therefore the point is $(9, \frac{5}{2})$

(d) This is an internal division. Hence, the positive values of the ratio must be maintained

$$x = \frac{px_2 + qx_1}{p + q}$$
$$= \frac{(2 \times -2) + (3 \times -3)}{2 + 3}$$
$$= \frac{-4 - 9}{5}$$
$$= \frac{-13}{5}$$

$$y = \frac{py_2 + qy_1}{p + q}$$
$$= \frac{(2 \times 5) + (3 \times (-1))}{2 + 3}$$
$$= \frac{10 - 3}{5}$$
$$= \frac{7}{5}$$

Therefore the point is $(-\frac{13}{5}, \frac{7}{5})$

(e) This is an external division. In the ratio 3:5, none of the values is negative. However, for an external division, one of the ratio values has to be negative. Hence, we can make one of the ratio values to be negative by taking the ratio, 3 : −5. We now proceed as follows:

$$x = \frac{px_2 + qx_1}{p + q}$$

$$= \frac{(3 \times 2) + (-5 \times 4)}{3 + (-5)}$$

$$= \frac{6 - 20}{-2}$$

$$= \frac{-14}{-2}$$

$$= 7$$

$$y = \frac{py_2 + qy_1}{p + q}$$

$$= \frac{(3 \times (-6)) + (-5 \times 2)}{3 + (-5)}$$

$$= \frac{-18 - 10}{-2}$$

$$= \frac{-28}{-2}$$

$$= 14$$

Therefore the point is (7, 14)

6. Determine the ratio in which the point (−11, 16) divides the line joining the points A(−1, 2) and B(4, −5)

Solution

Let the ratio be p : q. Then the x coordinate of the point which divides the line AB is given by:

$$x = \frac{px_2 + qx_1}{p + q}$$

Substituting −11 for x, −1 for x_1 and 4 for x_2 gives:

$$-11 = \frac{p(4) + q(-1)}{p + q}$$

$$-11 = \frac{4p - q}{p + q}$$

−11(p + q) = 4p − q

−11p − 11q = 4p − q

−11p − 4p = 11q − q

−15p = 10q

Divide both sides by −15. This gives

$$p = \frac{10q}{-15}$$

Now divide both sides by q in order to get the ratio p : q. This gives:

$\frac{p}{q} = -\frac{2}{3}$ (In its lowest term)

Hence, p : q = –2 : 3

Therefore, the point (–11, 16) divides the line AB externally in the ration 2 : 3.

Note that the word "externally'" means that the negative sign in the ratio has to be removed.

Note that this question can also be solved by using the y coordinate of the point that divides the line AB. In this case the equation to use will be:

$$y = \frac{py_2 + qy_1}{p + q}$$

Substituting 16 for y, 2 for y_1 and –5 for y_2 will still give the ratio p : q = –2 : 3

7. Determine the ratio in which the point (16, –19) divides the line joining the points A(–1, 2) and B(4, –5)

Solution

Let the ratio be p : q. Then the y coordinate of the point which divides the line AB is given by:

$$y = \frac{py_2 + qy_1}{p + q}$$

Substituting –19 for y, 2 for y_1 and –5 for y_2 gives:

$$-19 = \frac{p(-5) + q(2)}{p + q}$$

$$-19 = \frac{-5p + 2q}{p + q}$$

–19(p + q) = –5p + 2q

–19p – 19q = –5p + 2q

–19p + 5p = 19q + 2q

–14p = 21q

Divide both sides by –14. This gives

$$p = \frac{21q}{-14}$$

Now divide both sides by q in order to get the ratio p : q. This gives:

$\frac{p}{q} = -\frac{3}{2}$ (In its lowest term)

Hence, p : q = –3 : 2

Therefore, the point (16, –19) divides the line AB externally in the ration 3 : 2.

15

8. Determine the ratio in which the x-axis divides the line joining the points (2, 3) and (5, –4).

Solution

On the x-axis the value of y is 0 (zero).

Let the ratio be p : q. Then the y coordinate of the point which divides the line is given by:

$$y = \frac{py_2 + qy_1}{p + q}$$

Substitute 0 for y, 3 for y_1 and –4 for y_2. This gives:

$$0 = \frac{p(-4) + q(3)}{p + q}$$

$$0 = \frac{-4p + 3q}{p + q}$$

$$0(p + q) = -4p + 3q$$
$$0 = -4p + 3q$$
$$4p = 3q$$

Divide both sides by 4. This gives

$$p = \frac{3q}{4}$$

Now divide both sides by q in order to get the ratio p : q. This gives:

$$\frac{p}{q} = \frac{3}{4}$$

Hence, p : q = 3 : 4

Therefore, the x-axis divides the line internally in the ration 3 : 4.

9. Find the ratio in which the line $x - y = 0$ divides the line LM joining the points L(–2, 4) and M(3, –1).

Solution

The line $x - y = 0$ shows that $x = y$ (when rearranged). This means that the x and y coordinates of the point which divides the line LM are equal. Hence, if we take the ratio of division to be p : q, then it follows that:

$$x = \frac{px_2 + qx_1}{p + q} = y = \frac{py_2 + qy_1}{p + q} \qquad \text{(since } x = y\text{)}$$

Therefore: $\dfrac{px_2 + qx_1}{p + q} = \dfrac{py_2 + qy_1}{p + q}$

Multiplying both sides of the equation by p + q simplifies the equation to give:

$$px_2 + qx_1 = py_2 + qy_1$$

Substitute $x_1 = -2$, $x_2 = 3$, $y_1 = 4$ and $y_2 = -1$. This gives:

$p(3) + q(-2) = p(-1) + q(4)$

$3p - 2q = 4q - p$

$3p + p = 4q + 2q$

$4p = 6q$

$\dfrac{p}{q} = \dfrac{6}{4}$

(Take note of this simplification which means you divide by 4q, similar to other examples)

Hence, $\dfrac{p}{q} = \dfrac{3}{2}$ (In its lowest term)

Or, $p : q = 3 : 2$

Therefore the line $x - y = 0$ divides the line LM internally in the ratio $3 : 2$

10. Find the distance between the point (2, −3) and the line $3x - 4y + 7$.

<u>Solution</u>

Comparing the line $3x - 4y + 7$ with the standard form $ax + by + c = 0$ shows that :

a = 3, b = −4 and c = 7.

The point (2, −3) shows that $x_1 = 2$ and $y_1 = -3$.

Hence, the distance of the point from the line is given by:

$d = \dfrac{|ax_1 + by_1 + c|}{\sqrt{a^2 + b^2}}$

$= \dfrac{|3(2) + -4(-3) + 7|}{\sqrt{3^2 + (-4)^2}}$

$= \dfrac{|6 + 12 + 7|}{\sqrt{9 + 16}}$

$= \dfrac{|25|}{\sqrt{25}}$

$= \dfrac{25}{5}$

= 5 units

11. Find the perpendicular distance between the line $2y = 5x - 11$ and (−8, −1)

<u>Solution</u>

$2y = 5x - 11$

Or, $5x - 2y - 11 = 0$

Comparing this line with the standard form $ax + by + c = 0$ shows that :

a = 5, b = –2 and c = –11.

The point (–8, –1) shows that x_1 = –8 and y_1 = –1.

Hence, the distance of the point from the line is given by:

$$d = \frac{|ax_1 + by_1 + c|}{\sqrt{a^2 + b^2}}$$

$$= \frac{|5(-8) + -2(-1) - 11|}{\sqrt{5^2 + (-2)^2}}$$

$$= \frac{|-40 + 2 - 11|}{\sqrt{25 + 4}}$$

$$= \frac{|-49|}{\sqrt{29}}$$

$$= \frac{49}{5.3852} \qquad \text{(Ignore the negative sign as indicated by the bar lines)}$$

$$= 9.1 \text{ units}$$

12. Two parallel lines have equations $2x - y - 5 = 0$ and $y - 2x - 3 = 0$. Find the distance between the two lines.

Solution

Let us find a point on the line $2x - y - 5 = 0$. Using a value of x = 1, let us find the corresponding value of y as follows:

$2x - y - 5 = 0$

$2(1) - y - 5 = 0$

$y = 2 - 5$

$y = -3$

Therefore, a point on the line $2x - y - 5 = 0$ is (1, –3). The distance between this point and the second line gives the distance between the two lines.

The equation of the second line is $y - 2x - 3 = 0$. This can also be expressed as:

$-2x + y - 3 = 0$.

Comparing this line with the standard form $ax + by + c = 0$ shows that :

a = –2, b = 1 and c = –3.

The point (1, –3) shows that x_1 = 1 and y_1 = –3.

Hence, the distance of the point from the line is given by:

$$d = \frac{|ax_1 + by_1 + c|}{\sqrt{a^2 + b^2}}$$

$$= \frac{|-2(1) + 1(-3) - 3|}{\sqrt{(-2)^2 + (1)^2}}$$

$$= \frac{|-2-3-3|}{\sqrt{4+1}}$$

$$= \frac{|-8|}{\sqrt{5}}$$

$$= \frac{8}{\sqrt{5}} \quad \text{(Ignore the negative sign as indicated by the bar lines)}$$

$$= 3.58$$

Therefore, the distance between the two lines is 3.58 units

13. Find the gradient of the line joining the points (4, 5) and (2, 3), and the angle of slope of the line.

Solution

The gradient of a line is given by:

$$m = \frac{y_2 - y_1}{x_2 - x_1}$$

$$= \frac{3-5}{2-4}$$

$$= \frac{-2}{-2}$$

$$m = 1$$

The angle of slope of the line is given by:

$$\theta = \tan^{-1}m$$

$$= \tan^{-1}1$$

$$\theta = 45^\circ$$

14. Find the slope of the line joining the points (−1, 4) and (1, 3). Hence, determine the angle of slope of the line.

Solution

The gradient of a line is given by:

$$m = \frac{y_2 - y_1}{x_2 - x_1}$$

$$= \frac{3-4}{1-(-1)}$$

$$= \frac{-1}{2}$$

$$= -\frac{1}{2}$$

$$m = -0.5$$

The angle of slope of the line is given by:

$$\theta = \tan^{-1}m$$

$$= \tan^{-1}(-0.5)$$

$$= -26.6 \quad \text{(Note that } \tan^{-1}0.5 = 26.6°)$$

$$= -26.6 + 180$$

$$= 153.4°$$

This shows that when the slope of a line is negative, the angle of slope is an obtuse angle (i.e. angle between $90°$ and $180°$)

15. Determine if the line WX is parallel or perpendicular to line YZ in each of the following:

(a) W(2, 1), X(4, 5) and Y(0, 4), Z(−3, 1)

(b) W(−1, −3), X(3, 5) and Y(1, 7), Z(3, 6)

(c) W(−4, −3), X(−2, 2) and Y(1, −5), Z(6, −7)

(d) W(−3, 8), X(5, −2) and Y(7, −4), Z(11, −9)

Solution

(a) Let the slope of line WX be m_1.

$$m_1 = \frac{y_2 - y_1}{x_2 - x_1}$$

$$= \frac{3 - 1}{4 - 2}$$

$$= \frac{2}{2}$$

$$m_1 = 1$$

Similarly, let the gradient of YZ be m_2.

$$m_2 = \frac{y_2 - y_1}{x_2 - x_1}$$

$$= \frac{1 - 4}{-3 - 0}$$

$$= \frac{-3}{-3}$$

$$m_2 = 1$$

Since $m_1 = m_2$, (i.e. 1), it shows that the line WX and YZ are parallel.

(b) Let the slope of line WX be m_1.

$$m_1 = \frac{y_2 - y_1}{x_2 - x_1}$$

$$= \frac{5 - (-3)}{3 - (-1)}$$

$$= \frac{5 + 3}{3 + 1}$$

$$= \frac{8}{4}$$

$$m_1 = 2$$

Similarly, let the gradient of YZ be m_2.

$$m_2 = \frac{y_2 - y_1}{x_2 - x_1}$$

$$= \frac{6 - 7}{3 - 1}$$

$$m_2 = \frac{-1}{2}$$

Hence m_1 and m_2 are not equal. This shows that the two lines are not parallel.

However, $m_1 \times m_2 = 2 \times \dfrac{-1}{2}$

$$= -1$$

Therefore, $m_1 m_2 = -1$. This shows that WX and YZ are perpendicular lines.

(c) $m_1 = \dfrac{2-(-3)}{-2-(-4)}$

$$= \frac{2 + 3}{-2 + 4}$$

$$m_1 = \frac{5}{2}$$

$$m_2 = \frac{-7-(-5)}{6 - 1}$$

$$= \frac{-7+5}{5}$$

$$m_2 = \frac{-2}{5}$$

Hence, $m_1 \times m_2 = \dfrac{5}{2} \times \dfrac{-2}{5}$

$$= -1$$

This shows that WX and YZ are perpendicular lines.

(d) $m_1 = \dfrac{-2-8}{5-(-3)}$

$$= \frac{-10}{8}$$

$$m_1 = \frac{-5}{4}$$

$$m_2 = \frac{-9-(-4)}{11 - 7}$$

$$= \frac{-9 + 4}{4}$$

$$m_2 = \frac{-5}{4}$$

Since $m_1 = m_2$, it shows that the two lines are parallel.

16. Find the slopes of the following lines:

(a) 2x + 3y = 6

(b) $5x - y + 2 = 0$

(c) $\frac{1}{2}y - 3x = 1$

Solution

(a) Let us represent the equation in the gradient intercept form by making y the subject of the formula. This is obtained as follows:

$$2x + 3y = 6$$
$$3y = -2x + 6$$

Dividing each term by 3 gives y as follows:

$$y = -\frac{2}{3}x + 2$$

This is similar to $y = mx + c$, where m is the gradient. Comparing it with the equation above shows that: $m = -\frac{2}{3}$

Therefore, the slope/gradient of the line is $-\frac{2}{3}$

(b) $5x - y + 2 = 0$

 $5x + 2 = y$

Or, $y = 5x + 2$

When this is compared with $y = mx + c$, it shows that m = 5

Therefore, the slope of the line is 5.

(c) $\frac{1}{2}y - 3x = 1$

 $\frac{1}{2}y = 3x + 1$

Multiplying each term by 2 in order to clear out the fraction gives:

 $y = 6x + 2$

Comparing with $y = mx + c$ shows that:

 slope, m = 6

17. Determine the length of the perpendicular drawn from the point (4, −3) to the line $2x - 5y + 3 = 0$

Solution

The perpendicular distance between the point (x_1, y_1) and the line $ax + by + c = 0$, is given by:

$$D = \frac{|ax_1 + by_1 + c|}{\sqrt{a^2 + b^2}}$$

With (4, −3) as (x_1, y_1), and a = 2 and b = −5, we obtain the distance as follows:

$$D = \frac{|ax_1 + by_1 + c|}{\sqrt{a^2 + b^2}}$$

$$= \frac{|2 \times 4 + (-5 \times -3) + 3|}{\sqrt{2^2 + (-5)^2}}$$

$$= \frac{|8 + 15 + 3|}{\sqrt{4 + 25}}$$

$$= \frac{26}{\sqrt{29}} \text{ units}$$

18. Find the distance between the point (–4, –3) and the line $-3x - 4y + 1 = 0$

Solution

$$D = \frac{|ax_1 + by_1 + c|}{\sqrt{a^2 + b^2}}$$

With (–4, –3) as (x_1, y_1), and a = –3 and b = –4, we obtain the distance as follows:

$$D = \frac{|ax_1 + by_1 + c|}{\sqrt{a^2 + b^2}}$$

$$= \frac{|(-3 \times -4) + (-4 \times -3) + 1|}{\sqrt{(-3)^2 + (-4)^2}}$$

$$= \frac{|12 + 12 + 1|}{\sqrt{9 + 16}}$$

$$= \frac{25}{\sqrt{25}}$$

$$= \frac{25}{5}$$

$$D = 5 \text{ units}$$

19. Find the acute angle between the lines $x - y - 4 = 0$ and $3x + 5y = 8$

Solution

The slope of the first line is m_1. It is obtained as follows:

$x - y - 4 = 0$

$x - 4 = y$

Hence, y = $x - 4$

Comparing this equation with y = $mx + c$ shows that the slope, m = 1

Hence, $m_1 = 1$

Similarly, The slope of the second line is m_2. It is obtained as follows:

$3x + 5y = 8$

$5y = -3x + 8$

$y = \frac{-3}{5}x + \frac{8}{5}$

Hence, the slope, m = $-\frac{3}{5}$

Hence, $m_2 = -\frac{3}{5}$

Therefore, the acute angle between the two lines is given by:

$$\tan \theta = \left| \frac{m_2 - m_1}{1 + m_1 m_2} \right|$$

$$= \left| \frac{-\frac{3}{5} - 1}{1 + (-\frac{3}{5} \times 1)} \right|$$

$$= \left| \frac{-\frac{8}{5}}{1 - \frac{3}{5}} \right|$$

$$= \left| \frac{-\frac{8}{5}}{\frac{2}{5}} \right|$$

$$= \left| \frac{-8}{5} \times \frac{5}{2} \right|$$

$$= |-4|$$

$$\tan \theta = 4 \qquad \text{(Note that } |-a| = a)$$

$$\theta = \tan^{-1} 4$$

$$= 75.96°$$

Therefore, the acute angle between the lines is $75.96°$.

20. Find the obtuse angle between the lines $2x - \frac{1}{2} y = 3$ and $5y - 2x + 11 = 0$

Solution

From the first line, m_1 is obtained as follows:

$$2x - \frac{1}{2} y = 3$$

$$2x - 3 = \frac{1}{2} y$$

Multiplying each term by 2 to clear the fraction gives:

$$2(2x) - 2(3) = 2(\frac{1}{2} y)$$

$$4x - 6 = y$$

Hence, $y = 4x - 6$

Therefore, the slope, $m_1 = 4$

Similarly, the slope, m_2 of the second line is obtained as follows:

$$5y - 2x + 11 = 0$$

$$5y = 2x - 11$$

$$y = \frac{2}{5} x - \frac{11}{5}$$

Hence, $m_2 = \frac{2}{5}$

Therefore, the acute angle between the two lines is given by:

$$\tan \theta = \left| \frac{m_2 - m_1}{1 + m_1 m_2} \right|$$

$$= \left| \frac{\frac{2}{5} - 4}{1 + (\frac{2}{5} \times 4)} \right|$$

$$= \left| \frac{\frac{2-20}{5}}{1 + \frac{8}{5}} \right|$$

$$= \left| \frac{\frac{-18}{5}}{\frac{13}{5}} \right|$$

$$= \left| \frac{-18}{5} \times \frac{5}{13} \right|$$

$$= \left| \frac{-18}{13} \right|$$

$$\tan \theta = \frac{18}{13}$$

$$\theta = \tan^{-1} \frac{18}{13}$$

$$= \tan^{-1} 1.385$$

$$\theta = 54.2° \quad \text{(This is the acute angle)}$$

Therefore, the obtuse angle between the lines is given by:

$$180 - 54.2 = 125.8°$$

21. Find the angle between the lines $2x - 4y = 5$ and $6x - 3y = 8$

Solution

From the first line, m_1 is obtained as follows:

$$2x - 4y = 5$$
$$-4y = -2x + 5$$
$$\frac{-4y}{-4} = \frac{-2x}{-4} + \frac{5}{-4}$$
$$y = \frac{1}{2}x - \frac{5}{4}$$

Therefore, the slope, $m_1 = \frac{1}{2}$

Similarly, the slope, m_2 of the second line is obtained as follows:

$$6x - 3y = 8$$
$$-3y = -6x + 8$$
$$y = \frac{-6}{-3}x + \frac{8}{-3}$$
$$y = 2x - \frac{8}{3}$$

Hence, $m_2 = 2$

Therefore, the acute angle between the two lines is given by:

25

$$\tan\theta = \left|\frac{m_2 - m_1}{1 + m_1 m_2}\right|$$

$$= \left|\frac{2 - \frac{1}{2}}{1 + (2 \times \frac{1}{2})}\right|$$

$$= \left|\frac{\frac{3}{2}}{1 + 1}\right|$$

$$= \left|\frac{\frac{3}{2}}{2}\right|$$

$$= \left|\frac{3}{2} \times \frac{1}{2}\right|$$

$$\tan\theta = \frac{3}{4}$$

$$\theta = \tan^{-1} 0.75$$

$$\theta = 36.9^\circ$$

The acute angle between the two line is 36.9°, while the obtuse angle between the two lines is:

$$180 - 36.9 = 143.1^\circ$$

22. Find the equations of the following lines:

(a) slope: –3, intercept on the y-axis: 2

(b) slope: $\frac{5}{2}$, intercept on the y-axis: –4

Solution

(a) The equation of a line in the gradient-intercept form is given by:

$y = mx + c$

Substituting the given values gives the equation of the line as follows:

$y = mx + c$

$y = -3x + 2$ (Since m = –3 and c = 2)

Hence, the equation of the line is $y = -3x + 2$

(b) $m = \frac{5}{2}$ and c = –4

$y = mx + c$

$y = \frac{5}{2}x - 4$

Multiply each term by 2 to clear the fraction as follows:

$2(y) = 2(\frac{5}{2}x) - 2(4)$

$2y = 5x - 8$

Therefore, the equation of the line is $2y = 5x - 8$

23. Find the equations of the following lines:

(a) slope: $-\frac{1}{2}$, intercept on the x-axis: 3

(b) slope: 2, intercept on the x-axis: –5

Solution

(a) The intercept on the x-axis occurs at the point where y = 0. Hence, an x-axis intercept above can be expressed as (3, 0). We now have three points given by:

$$x = 3, \ y = 0, \ m = -\frac{1}{2}$$

We now substitute each of these values into the gradient-intercept form of a line as follows:

$$y = mx + c$$
$$0 = (-\frac{1}{2} \times 3) + c$$
$$0 = -\frac{3}{2} + c$$
$$\frac{3}{2} = c$$
$$c = \frac{3}{2}$$

We can now substitute the values of m and c into the gradient-intercept form in order to obtain the equation of the line as follows:

$$y = mx + c$$
$$y = -\frac{1}{2}x + \frac{3}{2}$$

Multiply each term by 2 in order to clear the fraction

$$2(y) = 2(-\frac{1}{2}x) + 2(\frac{3}{2})$$
$$2y = -x + 3$$

Or $\quad x + 2y = 3$

The equation of the line is $x + 2y = 3$

(b) The x-axis intercept is (–5, 0). Hence:

$$x = -5, \ y = 0, \ m = 2$$

We now substitute each of these values into the equation below in order to obtain c.

$$y = mx + c$$
$$0 = 2(-5) + c$$
$$0 = -10 + c$$
$$c = 10$$

With m = 2, and c = 10, the equation of the live is given by:

$$y = mx + c$$
$$y = 2x + 10$$

Therefore, the equation of the line is $y = 2x + 10$

24. Find the x and y axes intercepts of the lines below:

(a) $3x - 5y = 10$

(b) $2x + y - 12 = 0$

Solution

Substitute $y = 0$ in the equation and solve for x in order to obtain the x-intercept. This gives:

$3x - 5y = 10$

$3x - 5(0) = 10$

$3x = 10$

$x = \dfrac{10}{3}$

Therefore, the x-intercept is $\dfrac{10}{3}$ or $(\dfrac{10}{3}, 0)$

Similarly, put $x = 0$ in the equation and solve for y in order to obtain the y-intercept. This gives:

$3x - 5y = 10$

$3(0) - 5y = 10$

$-5y = 10$

$y = \dfrac{10}{-5}$

$y = -2$

Therefore, the y-intercept is -2 or $(0, -2)$

(b) When $y = 0$, we obtain the x-intercept as follows:

$2x + y - 12 = 0$

$2x + 0 - 12 = 0$

$2x = 12$

$x = \dfrac{12}{2}$

$x = 6$

Therefore, the x-intercept is 6 or $(6, 0)$

Similarly, put $x = 0$ in the equation and solve for y in order to obtain the y-intercept. This gives:

$2x + y - 12 = 0$

$2(0) + y - 12 = 0$

$y = 12$

Therefore, the y-intercept is 12 or $(0, 12)$

25. Find the equation of the line which passes through the point $(1, -3)$ and is parallel to the line $4x - 2y = 5$

Solution

We first find the gradient of the line $4x - 2y = 5$ as follows:

$4x - 2y = 5$

$$4x - 5 = 2y$$
$$y = \frac{4}{2}x - \frac{5}{2}$$
$$y = 2x - \frac{5}{2}$$

Hence, the gradient of this line is 2 (when compared to $y = mx + c$)

Since the required line is parallel to this line, then their gradient are equal. Hence, the gradient of the required line is 2. Recall that the line passes through (1, −3). This point represents (x_1, y_1). Therefore, the equation of the line is given by:

$$m = \frac{y - y_1}{x - x_1}$$

Or, $m(x - x_1) = y - y_1$

Substitute, $m = 2$, $x_1 = 1$ and $y_1 = -3$ into the equation above. This gives:

$$m(x - x_1) = y - y_1$$
$$2(x - 1) = y - (-3)$$
$$2x - 2 = y + 3$$
$$2x - y = 3 + 2$$
$$2x - y = 5$$

Therefore, the equation of the line is $2x - y = 5$

26. Find the equation of the line which passes through the point (−2, −5) and is parallel to the line $2x - 3y - 1 = 0$

Solution

$$2x - 3y - 1 = 0$$
$$-3y = -2x + 1$$
$$y = \frac{-2x}{-3} + \frac{1}{-3}$$
$$y = \frac{2}{3}x - \frac{1}{3}$$

Comparing this line with $y = mx + c$, shows that the gradient of the line is $\frac{2}{3}$

Hence, the gradient of the required line is $\frac{2}{3}$ since they are parallel lines.

Therefore, $m = \frac{2}{3}$

Let the equation of the required line be

$$y = mx + c$$

Since the line passes through (−2, −5), it means that we can substitute $m = \frac{2}{3}$, $x = -2$ and $y = -5$ into the line: $y = mx + c$. This gives:

$$-5 = \frac{2}{3}(-2) + c$$
$$-5 = -\frac{4}{3} + c$$
$$-5 + \frac{4}{3} = c$$
$$\frac{-15 + 4}{3} = c$$

$$c = \frac{-11}{3}$$

We now substitute $c = \frac{-11}{3}$ and $m = \frac{2}{3}$ into the line $y = mx + c$, as follows:

$$y = \frac{2}{3}x - \frac{11}{3}$$

Multiply each term by 3 to clear the fraction.

$$3(y) = 3\left(\frac{2}{3}x\right) - 3\left(\frac{11}{3}\right)$$

$$3y = 2x - 11$$

Or, $2x - 3y - 11 = 0$

Therefore the equation of the line is $2x - 3y - 11 = 0$

Note that the method applied in example 18 is slightly different from the method used here. Any of the two methods can be applied in problems like these.

27. Find the equation of the line which passes through the point $(-1, 2)$ and is perpendicular to the line $x - 4y - 2 = 0$

<u>Solution</u>

$$x - 4y - 2 = 0$$

$$-4y = -x + 2$$

$$y = \frac{-x}{-4} + \frac{2}{-4}$$

$$y = \frac{1}{4}x - \frac{1}{2}$$

Hence, the gradient of the line is $\frac{1}{4}$

Let the gradient of the required line be m. Therefore, since they are perpendicular lines, the product of their gradient is −1.

Therefore, $\frac{1}{4}(m) = -1$

$$\frac{m}{4} = -1$$

$$m = 4(-1)$$

$$m = -4$$

Hence, the gradient of the required line is −4

Let us now substitute the point $(-1, 2)$ and the gradient m = −4 into the equation $y = mx + c$ in order to obtain the value of c as follows:

$$y = mx + c$$

$$2 = -4(-1) + c$$

$$2 = 4 + c$$

$$c = 2 - 4$$

$$c = -2$$

We now substitute c = −2 and m = −4 into the line $y = mx + c$ as follows:

$$y = -4x - 2$$

Or, $4x + y + 2 = 0$

Therefore, the equation of the line is $4x + y + 2 = 0$

28. Find the equation of the line which passes through the point $(3, -2)$ and is perpendicular to the line $5x + 2y - 4 = 0$

Solution

$$5x + 2y - 4 = 0$$
$$2y = -5x + 4$$
$$y = \frac{-5x}{2} + \frac{4}{2}$$
$$y = -\frac{5}{2}x + 2$$

Hence, the gradient of the line is $-\frac{5}{2}$

Let the gradient of the required line be m. Therefore, since they are perpendicular lines, the product of their gradient is -1.

Therefore, $-\frac{5}{2}(m) = -1$

$$\frac{-5m}{2} = -1$$
$$-5m = 2(-1)$$
$$m = \frac{-2}{-5}$$
$$m = \frac{2}{5}$$

Therefore, with the point $(3, -2)$ taken as (x_1, y_1), the equation of the required line is given by:

$$m = \frac{y - y_1}{x - x_1}$$

$$\frac{2}{5} = \frac{y - (-2)}{x - 3}$$

$$\frac{2}{5} = \frac{y + 2}{x - 3}$$

$$2(x - 3) = 5(y + 2)$$
$$2x - 6 = 5y + 10$$
$$2x - 5y - 16 = 0$$

The equation of the line is $2x - 5y - 16 = 0$

29. Find the point of intersection of the lines $3x - 5y = -1$ and $2x + 7y = -11$

Solution

In order to find the point of intersection of two lines, we simply solve the equations of the lines simultaneously as follows:

$$3x - 5y = -1 \dots\dots\dots(1)$$
$$2x + 7y = -11 \dots\dots\dots(2)$$

In order to make the coefficient of x to be the same (elimination method) in the two equations above, let us multiply equation (1) by 2 and equation (2) by 3. This gives:

$$2(3x - 5y = -1)$$
$$3(2x + 7y = -11)$$

Expanding the brackets gives equation (3) and (4) as follows:

$$6x - 10y = -2 \quad ....(3)$$
$$\underline{6x + 21y = -33 \quad ....(4)}$$

Equation (3) − Equation (4): $-31y = 31$

$$y = \frac{31}{-31}$$

$$y = -1$$

Substitute −1 for y in equation (1)

$3x - 5y = -1$ ....(1)

$3x - 5(-1) = -1$

$3x + 5 = -1$

$3x = -1 - 5$

$x = \dfrac{-6}{3}$

$x = -2$

Therefore the point of intersection of the lines is (−2, −1)

30. Find the point of intersection of the lines $5x - y = -17$ and $2x - 3y = 1$

<u>Solution</u>

In order to find the point of intersection of two lines, we simply solve the equations of the lines simultaneously as follows:

$5x - y = -17$ ....(1)

$2x - 3y = 1$ ....(2)

From equation (1): $y = 5x + 17$ ....(3)

Substitute $5x + 17$ for y in equation (2)

$2x - 3y = 1$ ....(2)

$2x - 3(5x + 17) = 1$

$2x - 15x - 51 = 1$

$-13x = 1 + 51$

$x = \dfrac{52}{-13}$

$x = -4$

Substitute −4 for x in equation (3)

$y = 5x + 17$ ....(3)

$y = 5(-4) + 17$

$y = -20 + 17$

$y = -3$

Therefore the point of intersection of the lines is (−4, −3)

31. Find the equation of the line which is parallel to the line $3x + 4y = 12$ and makes an intercept of 5 units on the x-axis.

<u>Solution</u>

$3x + 4y = 12$

$4y = -3x + 12$

$y = \dfrac{-3}{4}x + 3$ (After dividing each term by 4)

Hence, the gradient of this line is $-\dfrac{3}{4}$

Therefore the gradient of the required line is also $-\dfrac{3}{4}$ since they are parallel lines. Recall that from the question, the required line passes through the x-axis at the point $x = 5$. At any point of interception on the x-axis, $y = 0$. Hence, the required line passes through (5, 0). Using this point as (x_1, y_1), we now obtain the required line by using the formula:

$m = \dfrac{y - y_1}{x - x_1}$

Or, $m(x - x_1) = y - y_1$

$-\dfrac{3}{4}(x - 5) = y - 0$

$-3(x - 5) = 4y$

$-3x + 15 = 4y$

$3x + 4y - 15 = 0$

The equation of the line is $3x + 4y - 15 = 0$

32. A line passes through the points (2, –4) and (2, –1). Determine:
(a) the equation of the line
(b) the equation of the a second line which passes through (2, –1) and is perpendicular to the first line.

Solution

(a) The equation of the line is given by:

$\dfrac{y - y_1}{x - x_1} = \dfrac{y_2 - y_1}{x_2 - x_1}$

$\dfrac{y - (-4)}{x - 2} = \dfrac{-1 - (-4)}{2 - 2}$

$\dfrac{y + 4}{x - 2} = \dfrac{3}{0}$

$3(x - 2) = 0(y + 4)$

$3x - 6 = 0$

$x - 2 = 0$

(b) The equation of the line above (first line) is:

$x - 2 = 0$

It can also be represented as:

$0y + x - 2$

Or, $0y = -x + 2$

$y = \dfrac{-1}{0}x + \dfrac{2}{0}$

Therefore, $m_1 = \dfrac{-1}{0}$ (i.e. undefined)

Hence, $m_2 = 0$.

Note that if two lines are perpendicular and the gradient of one undefined, then the gradient of the other will be 0 (zero)

Therefore the equation of the second line is given by:

$$m_2 = \frac{y - y_1}{x - x_1}$$

$$0 = \frac{y-(-1)}{x - 2}$$ (Note that (x_1, y_1) is the point $(2, -1)$ from the question)

$$y + 1 = 0(x - 2)$$
$$y + 1 = 0$$

Or, $y = -1$

Therefore the equation of the second line is $y = -1$

33. Determine the equation of a line which is parallel to the line $x = -5$ and passing through the point $(-2, -4)$.

Solution

$x = -5$

This equation can also be represented by introducing y as follows:

$x = -5 + 0y$

$0y = x + 5$

$$y = \frac{1}{0}x + \frac{5}{0}$$

Hence, $m_1 = \dfrac{1}{0}$ (undefined)

Therefore, $m_2 = \dfrac{1}{0}$ (Since they are parallel lines)

Hence, the equation of the line through $(-2, -4)$ is given by:

$$m_2 = \frac{y - y_1}{x - x_1}$$

$$\frac{1}{0} = \frac{y-(-4)}{x-(-2)}$$

$$x + 2 = 0(y + 4)$$
$$x + 2 = 0$$

Or, $x = -2$

Therefore the equation of the line is $x = -2$

Questions 32 and 33 above shows that there is a direct method (without solving) to obtain equations of lines which are parallel to the x-axis or parallel to the y-axis. Also, the equation of a line perpendicular to either the x-axis or y-axis can be obtained directly without solving the problem. For example, if a line is given by $x = 5$, it means that the line is parallel to the y-axis. A second line which is parallel to $x = 5$ and passes through $(-4, 3)$ is simply obtained by equating x

with the x coordinate of the given point. Hence, the equation of the parallel line is $x = -4$. The equation of a line perpendicular to $x = 5$ and passing through the $(-4, 3)$ is simply obtained by equating y with the y coordinate of the given point. Hence, the equation of the perpendicular line is y = 3.

Similarly, if a line is given by y = –8, it means that the line is parallel to the x-axis. A second line which is parallel to y = –8 and passes through (1, –5) is simply obtained by equating y with the y coordinate of the given point. Hence, the equation of the parallel line is y = –5. The equation of a line perpendicular to y = –8 and passing through the (1, –5) is simply obtained by equating x with the x coordinate of the given point. Hence, the equation of the perpendicular line is $x = 1$.

Exercise 1

1. Find the gradients of the following lines passing through the points:
(a) P(1, –2) and Q(3, 2)
(b) (–1, 5), (3, 0)
(c) $(\frac{2}{5}, \frac{1}{4})$ and $(\frac{1}{5}, -\frac{1}{2})$
(d) (2.45, 0.12) and (–0.55, –1.48)
2. Two points on a line are (2, b) and (3, 5). If the gradient of the line is –1, find the value of b.
3. Find the distance between the following points:
(a) M(1, 2) and N(0, 3)
(b) S(5, 1) and B(2, –1)
(c) (–2, 7) and (–3, 2)
(d) (1, –2) and (–3, 1)
4. Find the mid-points of the lines that join the following points:
(a) X(2, 6) and Y(–2, 4)
(b) A(3, 3) and B(1, –1)
(c) (0, 9) and (2, –1)
(d) (–5, 1) and (7, 3)
5. Find the coordinates of the point which divide the lines joining the following points in the stated ratios.
(a) (2.5) and B(1, 0) in the ratio 3:1
(b) P(1.3) and Q(–2, 1) in the ratio 1: –2
(c) (3, 1) and (2, –2) in the ratio 2:1 externally
(d) (5, 2) and (1, –1) in the ratio 1:3 externally
(e) (2, 2) and (–1, 3) in the ratio 2:5 internally
6. Determine the ratio in which the point (0, 3) divides the line AB joining the points A(2, 1) and

B(−3, 6)

7. Determine the ratio in which the point (4, 2) divides the line joining points (4, −1) and (4, 3)

8. Determine the ratio in which the y-axis divides the line joining the points (2, −3) and (5, 6).

9. Find the ratio in which the line $2x + y = 0$ divides the line PQ joining the points P(2, −2) and Q(3, 7).

10. Two points P and Q divides line ST into three equal parts. If S and T are (2, 7) and (−4, −8) respectively, find the points P and Q.

11. Find the distance between the point (5, −1) and the line $2x + y − 4$.

12. Find the perpendicular distance between the line $3x = 2y − 5$ and (−4, 1)

13. Find the gradient of the line joining the points (1, −2) and (2, −5), and the angle of slope of the line.

14. Find the slope of the line joining the points (3, 2) and (5, −3). Hence, determine the angle of slope of the line.

15. Determine if the line AB is parallel or perpendicular to line BC in each of the following:

(a) A(4, 3), B(2, 1) and B(1, 5), C(−3, 1)

(b) A(2, 5), B(1, 3) and B(3, 8), C(5, 7)

(c) A(1, 2), B(2, 3) and B(2, −6), C(1, −5)

(d) A(7, 3), B(4, −2) and B(4, 11), C(9, 8)

16. Find the slopes of the following lines:

(a) $5x + 2y = 9$

(b) $9x − 3y + 5 = 0$

(c) $\frac{2}{5} y − 2x = 4$

17. Determine the length of the perpendicular drawn from the point (2, −2) to the line $x − 3y + 7 = 0$

18. Find the distance between the point (2, −5) and the line $5x − 2y − 11 = 0$

19. Find the acute angle between the lines $2x − y − 5 = 0$ and $5x − 3y = 5$

20. Find the obtuse angle between the lines $4x − \frac{2}{3} y = 1$ and $y + x + 5 = 0$

21. Find the angle between the lines $x − 3y = 7$ and $2x − 5y = 2$

22. Find the equations of the following lines:

(a) slope: 1, intercept on the y-axis: 5

(b) slope: $\frac{1}{2}$, intercept on the y-axis: 1

23. Find the equations of the following lines:

(a) slope: −3, intercept on the x-axis: 7

(b) slope: −1, intercept on the x-axis: 10

24. Find the x and y axes intercepts of the lines below:

(a) $2x − 7y = 3$

(b) $5x + y - 8 = 0$

25. Find the equation of the line which passes through the point (1, −3) and is parallel to the line $3x - 5y = 1$

26. Find the equation of the line which passes through the point (1, −1) and is parallel to the line $x - y - 3 = 0$

27. Find the equation of the line which passes through the point (−5, 1) and is perpendicular to the line $2x - y - 5 = 0$

28. Find the equation of the line which passes through the point (7, −5) and is perpendicular to the line $x + 3y - 2 = 0$

29. Find the point of intersection of the lines $3x - 5y = 8$ and $2x + 7y = -5$

30. Find the point of intersection of the lines $x - y = -1$ and $3x - y = 3$

31. Find the equation of the line which is parallel to the line $2x + y = 6$ and makes an intercept of 10 units on the x-axis.

32. A line passes through the points (1, 1) and (5, −1). Determine:

(a) the equation of the line

(b) the equation of the a second line which passes through (3, −5) and is perpendicular to the first line.

33. Determine the equation of a line which is parallel to the line $2x = -7$ and passing through the point (1, −2).

34. Two parallel lines are given by $2x + y = -7$ and $2y + 4x = 5$. Find the distance between the lines.

35. Two parallel lines are given by $x - 2y = 1$ and $2y - x = 12$. Find the perpendicular distance between the lines.

CHAPTER 2
AREAS OF TRIANGLE AND QUADRILATERAL USING THE X-Y COORDINATES

Area of Triangle

The area of a triangle ABC when given the coordinates of its three vertices is given by:

$$\text{Area} = \left| \frac{Ax(By-Cy) + Bx(Cy-Ay) + Cx(Ay-By)}{2} \right|$$

where Ax, Ay, Bx, By and Cx, Cy are the x and y coordinates of the points A, B and C respectively.

Area of Quadrilateral

The area of a quadrilateral ABCD when given the coordinates of its four vertices is given by:

$$\text{Area} = \left| \frac{Ax(By-Dy) + Bx(Cy-Ay) + Cx(Dy-By) + Dx(Ay-Cy)}{2} \right|$$

where Ax, Ay, Bx, By, Cx, Cy, Dx and Dy are the x and y coordinates of the points A, B, C, and D respectively.

Examples

1. Three points A(7, 9), B(3, 5) and C(5, 1), are the vertices of a triangle ABC. If P and Q are the mid-points of AB and AC respectively, find:
(a) the equation of the line PQ
(b) E and F, the points of intersection of the line PQ and the x and y axes respectively
(c) the area of triangle ABC

<u>Solution</u>
(a) The mid-point of AB is given by:
$$\left(\frac{x_1 + x_2}{2}, \frac{y_1 + y_2}{2} \right)$$

$$= \left(\frac{7+3}{2}, \frac{9+5}{2} \right)$$

$$= \left(\frac{10}{2}, \frac{14}{2} \right)$$

$$= (5, 7)$$

Hence, P is (5, 7)
The mid-point of AC is given by:
$$\left(\frac{7+5}{2}, \frac{9+1}{2} \right)$$

$$= \left(\frac{12}{2}, \frac{10}{2}\right)$$

$$= (6, 5)$$

Hence, Q is (5, 7)

The equation of the line PQ is given by:

$$\frac{y - y_1}{x - x_1} = \frac{y_2 - y_1}{x_2 - x_1} \quad \text{(where (5, 7) = } (x_1, y_1) \text{ and (6, 5) = } (x_2, y_2)\text{)}$$

$$\frac{y - 7}{x - 5} = \frac{5 - 7}{6 - 5}$$

$$\frac{y - 7}{x - 5} = \frac{-2}{1}$$

$$-2(x - 5) = y - 7$$

$$-2x + 10 = y - 7$$

$$y + 2x - 7 - 10 = 0$$

$$y + 2x - 17 = 0$$

The equation of line PQ is $y + 2x - 17 = 0$

(b) E is the point where PQ cuts the x-axis. At this point, y = 0. Hence, put y = 0 in the equation y + 2x − 17 = 0 in order to obtain x as follows:

$$y + 2x - 17 = 0$$

$$0 + 2x = 17$$

$$2x = 17$$

$$x = \frac{17}{2}$$

The point E is $\left(\frac{17}{2}, 0\right)$

Similarly, put x = 0 into y + 2x − 17 = 0 in order to obtain y. This will give the point F.

$$y + 2(0) - 17 = 0$$

$$y + 0 + 17 = 0$$

$$y = 17$$

Hence, F is the point (0, 17)

(c) The area of a triangle when given the coordinates of its three vertices is given by:

$$\text{Area} = \left| \frac{Ax(By - Cy) + Bx(Cy - Ay) + Cx(Ay - By)}{2} \right|$$

where Ax, Ay, Bx, By and Cx, Cy are the x and y coordinates of the points A, B and C respectively.

Therefore, with A(7, 9), B(3, 5) and C(5, 1), the area of the triangle is given by:

$$\text{Area} = \left| \frac{7(5 - 1) + 3(1 - 9) + 5(9 - 5)}{2} \right|$$

$$= \left| \frac{7(4) + 3(-8) + 5(4)}{2} \right|$$

$$= \left| \frac{28 - 24 + 20}{2} \right|$$

$$= \frac{24}{2}$$

$$= 12 \text{ square units}$$

2. The line y = 2x − 1 meets the y-axis at B and the line y = 5 at C
(a) Find the coordinates of B and C
(b) If A is the point (−4, 1), show that AB is perpendicular to BC
(c) Find the area of triangle ABC
<u>Solution</u>
(a) At the point where the line meets the y-axis, x = 0. Hence, put x = 0 in the equation y = 2x −
1 in order to get the value of y.

    y = 2x − 1
    y = 2(0) − 1
    y = 0 − 1
    y = −1
Therefore the coordinate of B is (0, −1)
In order to get point C, we solve the two equations simultaneously as follows:
    y = 2x − 1 ....(1)
    y = 5 ....(2)
Substitute 5 for y in equation (1)
    y = 2x − 1 ....(1)
    5 = 2x − 1
    5 + 1 = 2x
    $x = \dfrac{6}{2}$
    x = 3
The coordinates of C is (3, −1)

(b) A = (−4, 1) and B = (0, −1). Hence, the gradient of AB is given by:
$$m_1 = \frac{y_2 - y_1}{x_2 - x_1}$$

$$= \frac{-1 - 1}{0 - (-4)}$$

$$= \frac{-2}{4}$$

$$m_1 = -\frac{1}{2}$$
Recall that B = (0, −1) and C = (3, 5). Hence, the gradient of BC is given by:
$$m_2 = \frac{y_2 - y_1}{x_2 - x_1}$$

$$= \frac{5 - (-1)}{3 - 0}$$

$$= \frac{6}{3}$$

$m_2 = 2$

Let us find the the product of m_1 and m_2 as follows:

$$m_1 m_2 = -\frac{1}{2}(2)$$

$$= -1$$

Therefore, AB is perpendicular to BC since the product of their gradient is –1 (Recall that two lines are perpendicular if the product of their gradient is –1, i.e. $m_1 m_2 = -1$)

(c) A = (–4, 1), B = (0, –1)and C = (3, 5). Hence, the area of triangle ABC is given by:

$$\text{Area} = \left| \frac{Ax(By-Cy) + Bx(Cy-Ay) + Cx(Ay-By)}{2} \right|$$

$$\text{Area} = \left| \frac{-4(-1-5) + 0(5-1) + 3(1-(-1))}{2} \right|$$

$$= \left| \frac{-4(-6) + 0 + 3(2)}{2} \right|$$

$$= \left| \frac{24 + 6}{2} \right|$$

$$= \frac{30}{2}$$

$$= 15 \text{ square units}$$

3. Find the area of the quadrilateral whose vertices are A(–3, 1), B(5, –2), C(3, 4) and D(–4, –2)
Solution
The area of quadrilateral ABCD is given by:

$$\text{Area} = \left| \frac{Ax(By-Dy) + Bx(Cy-Ay) + Cx(Dy-By) + Dx(Ay-Cy)}{2} \right|$$

$$= \left| \frac{-3(-2-(-2)) + 5(4-1) + 3(-2-(-2)) + -4(1-4)}{2} \right|$$

$$= \left| \frac{-3(0) + 5(3) + 3(0) + -4(-3)}{2} \right|$$

$$= \left| \frac{0 + 15 + 0 + 12}{2} \right|$$

$$= \frac{27}{2}$$

$$= 13.5 \text{ square units}$$

4. Find the area of a quadrilateral whose vertices are P(–5, 3), Q(–2, 5), R(3, –2) and S(–3, 0)

<u>Solution</u>

The area of quadrilateral PQRS is given by:

$$\text{Area} = \left| \frac{Px(Qy-Sy) + Qx(Ry-Py) + Rx(Sy-Qy) + Sx(Py-Ry)}{2} \right|$$

$$= \left| \frac{-5(5-0) + -2(-2-3) + 3(0-5) + -3(3-(-2))}{2} \right|$$

$$= \left| \frac{-5(5) -2(-5) + 3(-5) -3(5)}{2} \right|$$

$$= \left| \frac{-25 + 10 - 15 - 15}{2} \right|$$

$$= \left| \frac{-45}{2} \right|$$

$$= |-22.5|$$

$$= 22.5 \text{ square units} \quad \text{(Ignore the negative sign as shown by the bar lines)}$$

5. The vertices A, B and C of a triangle ABC are (1, 2), (4, 7) and (6, −1) respectively. If M is the foot of the perpendicular from A to BC, determine:
(a) the coordinates of M
(b) the equation of AM
(c) the area of triangle ABC

<u>Solution</u>

The gradient of BC, i.e. (4, 7) and (6, −1) is given by:

$$\frac{-1-7}{6-4}$$

$$= \frac{-8}{2}$$

$$= -4$$

Any of the points b or C can be used to find the equation of BC. Using the point (4, 7) and the gradient of BC (i.e. −4), the equation of BC can be obtained as follows:

$$-4 = \frac{y-7}{x-4}$$

$$-4(x-4) = y-7$$

$$-4x + 16 = y - 7$$

$$4x + y = 23$$

Since the line AM is perpendicular to BC, then the gradient of AM is obtained by simply taking the negative inverse (negative reciprocal) of line BC

Hence, the gradient of AM = $\frac{-1}{-4} = \frac{1}{4}$ (From $m_1m_2 = -1$)

Therefore, using point A = (1, 2), we obtain the equation of AM as follows:

$$\frac{1}{4} = \frac{y-2}{x-1}$$

$(x-1) = 4(y-2)$

$x - 1 = 4y - 8$

$x - 4y = -7$

In order to obtain the coordinates of M, we solve the equation of line BC and AM simultaneously as follows:

$4x + y = 23$ ....(1)

$x - 4y = -7$ ....(2)

Equation (2) x 4 gives equation (3) as shown below. Then we use elimination method to equation (3) and (1) as follows:

$4x - 16y = -28$ ....(3)

$\underline{4x\ +\ y\ =\ 23}$ ....(1)

Equation: (1) – (3) $17y = 51$

$$y = \frac{51}{17}$$

$y = 3$

Substitute 3 for y in equation (2). This gives:

$x - 4y = -7$ ....(2)

$x - 4(3) = -7$

$x - 12 = -7$

$x = 12 - 7$

$x = 5$

Therefore:

(a) the coordinates of M is (5, 3)

(b) the equation of AM is $x - 4y = -7$ (as obtained above)

(c) A = (1, 2), B = (4, 7) and C = (6, –1)

The area of triangle ABC is given by:

$$\text{Area} = \left| \frac{Ax(By-Cy) + Bx(Cy-Ay) + Cx(Ay-By)}{2} \right|$$

$$\text{Area} = \left| \frac{1(7-(-1)) + 4(-1-2) + 6(2-7)}{2} \right|$$

$$= \left| \frac{1(8) + 4(-3) + 6(-5)}{2} \right|$$

$$= \left| \frac{8 - 12 - 30}{2} \right|$$

$$= \left| \frac{-34}{2} \right|$$

$$= |-17|$$

= 17 square units   (Ignore the negative sign as shown by the bar lines)

Exercise 2

1. Points X(6, 9), Y(2, 5) and Z(4, 3), are the vertices of a triangle XYZ. If A and B are the mid-points of XY and XZ respectively, find:

(a) the equation of the line AB

(b) P and Q, the points of intersection of the line AB and the x and y axes respectively

(c) the area of triangle XYZ

2. The line y = 5x − 3 meets the y-axis at M and the line y = 2 at N

(a) Find the coordinates of M and N

(b) If L is the point (−2, 1), find the area of triangle LMN

3. Find the area of the quadrilateral whose vertices are O(−5, 4), P(4, 5), Q(5, 2) and R(−2, −1)

4. Find the area of a quadrilateral whose vertices are A(−2, 4), B(2, 4), C(2, 0) and D(−2, 0)

5. The vertices P, Q and R of a triangle PQR are (1, −1), (3, 5) and (1, 4) respectively. If Z is the foot of the perpendicular from P to QR, determine:

(a) the coordinates of Z

(b) the equation of PZ

(c) the area of triangle PQR

6. Determine the area of the triangle formed by the positive x and y axes and the line 2x + y = 4

7. Line x + 2y = 3 meets the y-axis at A, while line x + y = 2 meets the x-axis at C. If the two lines intersect at B, find the area of quadrilateral ABCO, where O is the origin.

CHAPTER 3
LOCUS OF POINTS IN THE X-Y PLANE

What is a Locus

The movement of a path in a plane can be described relative to a set of points. A locus is the set of all points whose coordinates satisfy a given equation or condition.

Examples

1. Find the equation of the locus of points which is equidistant from points P(3, −2) and Q(4, 5).
Solution
Let a point on the locus be A(x, y). Since A is equidistant from P and Q, then it follows that the distance PA = QA
Hence, $|PA|^2 = |QA|^2$ (From length of a line)

$(x-3)^2 + (y - (-2))^2 = (x-4)^2 + (y-5)^2$

$x^2 - 6x + 9 + y^2 + 4y + 4 = x^2 - 8x + 16 + y^2 - 10y + 25$

$x^2 + y^2 - 6x + 4y + 13 = x^2 - y^2 - 8x - 10y + 41$

$x^2 + y^2 - x^2 - y^2 - 6x + 8x + 4y + 10y = 41 - 13$

$14y + 2x = 28$

Dividing each term by 2 gives:

$7y + x = 14$

2. Two points A and B have coordinates (−2, 3) and (−1, 4) respectively. M is a locus given by:

$|AM|^2 + |BM|^2 = 20.$

What is the equation of the locus M?
Solution
Let the coordinates of M be (x, y). Then we determine the equation of M as follows:

$|AM|^2 + |BM|^2 = 20$

$(x - (-2))^2 + (y - 3)^2 + (x - (-1))^2 + (y - 4)^2 = 20$

$x^2 + 4x + 4 + y^2 - 6y + 9 + x^2 + 2x + 1 + y^2 - 8y + 16 = 20$

$x^2 + x^2 + y^2 + y^2 + 4x + 2x - 6y - 8y + 4 + 9 + 1 + 16 = 20$

$2x^2 + 2y^2 + 6x - 14y + 30 - 20 = 0$

Dividing throughout by 2 give:

$x^2 + y^2 + 3x - 7y + 5 = 0$

Therefore, the equation of the locus M is $x^2 + y^2 + 3x - 7y + 5 = 0$

3. A point P moves in the x-y plane such that its distance from the point (1, −3) is equal to its distance from the line y − 4 = 0. Find the equation of the locus of the point P.
Solution
Let the coordinates of P be (x, y). Then the square of the distance between P and (1, −3) is given by:

$(x-1)^2 + (y - (-3))^2$ ....Equation (1)

Recall that the distance of a point from a line is given by:
$$\frac{|ax_1 + by_1 + c|}{\sqrt{a^2 + b^2}}$$

Hence, the distance of the point $P(x, y)$ from the line $y - 4 = 0$ is given by:
$$\frac{|0(x) + 1(y) - 4|}{\sqrt{0^2 + 1^2}} \qquad \text{(Note that a = 0 since there is no term in } x\text{)}$$
$$= \frac{|y - 4|}{1}$$
$$= |y - 4|$$

Squaring this distance gives:
$$(y - 4)^2 \dots\dots\dots\dots(2)$$

Since the distances between P and each of the two points are equal, then equation (1) and (2) above are equal. Equating them give:
$$(x - 1)^2 + (y - (-3))^2 = (y - 4)^2$$
$$(x - 1)^2 + (y + 3)^2 = (y - 4)^2$$
$$x^2 - 2x + 1 + y^2 + 6y + 9 = y^2 - 8y + 16$$
$$x^2 + y^2 - y^2 - 2x + 6y + 8y + 1 + 9 - 16 = 0$$
$$x^2 - 2x + 14y - 6 = 0$$

4. L is the point $(-2, 4)$ and M is the point $(3, -1)$. A variable point $Q (x, y)$ is such that $QL^2 - QM^2 = 10$. Find the equation of the locus Q.

Solution

Let the coordinates of Q be (x, y). Then we determine the equation of Q as follows:
$$|QL|^2 - |QM|^2 = 10$$
$$(x - (-2))^2 + (y - 4)^2 - (x - 3)^2 + (y - (-1))^2 = 10$$
$$(x + 2)^2 + (y - 4)^2 - [(x - 3)^2 + (y + 1)^2] = 10$$
$$x^2 + 4x + 4 + y^2 - 8y + 16 - [x^2 - 6x + 9 + y^2 + 2y + 1] = 10$$
$$x^2 + 4x + 4 + y^2 - 8y + 16 - x^2 + 6x - 9 - y^2 - 2y - 1 = 10$$
$$x^2 - x^2 + y^2 - y^2 + 4x + 6x - 8y - 2y + 4 + 16 - 9 - 1 = 10$$
$$10x - 10y + 10 - 10 = 0$$
$$10x - 10y = 0$$

Dividing throughout by 10 give:
$$x - y = 0$$

Therefore, the equation of the locus Q is $x - y = 0$

5. The line $x + 2y - 8 = 0$ meets the x and y axes at the points A and B respectively. If M is the midpoint of AB and P has coordinates $\left(\frac{11}{4}, -\frac{1}{2}\right)$, find the equation of the locus of all points in the plane whose distance from P is $|PM|$.

Solution
$$x + 2y - 8 = 0$$

On the x-axis, y = 0. Since the line meets the x-axis at the point A, we simply obtain the x coordinate of A by substituting 0 for y in the given equation. This gives:

$x + 2y - 8 = 0$

$x + 2(0) - 8 = 0$

$x = 8$

Hence, point A = (8, 0)

Since on the y-axis, $x = 0$, we can similarly obtain the y coordinate of B by substituting 0 for x in the given equation. This gives:

$x + 2y - 8 = 0$

$0 + 2y - 8 = 0$

$2y = 8$

$y = \dfrac{8}{2}$

$y = 4$

Hence, point B = (0, 4)

M, which is the midpoint of AB [A(8, 0) and B = (0, 4)] is obtained as follows:

$M = \left(\dfrac{x_1 + x_2}{2}, \; \dfrac{y_1 + y_2}{2} \right)$

$M = \left(\dfrac{8 + 0}{2}, \; \dfrac{0 + 4}{2} \right)$

$M = (4, 2)$

Let the required locus pass through the point L(x, y). From the question, the distance of the locus from P is $|PM|$. Hence, with P = $\left(\dfrac{11}{4}, \; -\dfrac{1}{2} \right)$, it follows that:

$|LP|^2 = |PM|^2$ (From distance between two points)

$(x - \dfrac{11}{4})^2 + (y - (-\dfrac{1}{2}))^2 \; = \; (\dfrac{11}{4} - 4)^2 + (-\dfrac{1}{2} - 2)^2$

$x^2 - \dfrac{11}{2}x + \dfrac{121}{16} + y^2 + y + \dfrac{1}{4} \; = \; \left(\dfrac{-5}{4}\right)^2 + \left(\dfrac{-5}{2}\right)^2$

$x^2 + y^2 - \dfrac{11}{2}x + y + \dfrac{121}{16} + \dfrac{1}{4} \; = \; \dfrac{25}{16} + \dfrac{25}{4}$

$x^2 + y^2 - \dfrac{11}{2}x + y + \dfrac{121}{16} + \dfrac{1}{4} - \dfrac{25}{16} - \dfrac{25}{4} = 0$

$x^2 + y^2 - \dfrac{11}{2}x + y + \dfrac{121 + 4 - 25 - 100}{16} = 0$

$x^2 + y^2 - \dfrac{11}{2}x + y + 0 = 0$

Multiplying each term by 2 to clear the fraction gives:

$2x^2 + 2y^2 - 11x + 2y = 0$

Therefore the equation of the locus is $2x^2 + 2y^2 - 11x + 2y = 0$

Exercise 3

1. Find the equation of the locus of points which is equidistant from points A(1, −5) and B(3, 2).

2. Two points M and N have coordinates (4, 5) and (3, −1) respectively. P is a locus given by:

$|MP|^2 + |NP|^2 = 12$.

What is the equation of the locus P?

3. A point E moves in the x-y plane such that its distance from the point (2, −1) is equal to its distance from the line $2y − 5 = 0$. Find the equation of the locus of the point E.

4. D is the point (1, 2) and P is point (−5, 3). A variable point B (x, y) is such that $BD^2 − BP^2 = 5$. Find the equation of the locus B.

5. The line $2x − 3y − 7 = 0$ meets the x and y axes at the points C and D respectively. If M is the midpoint of CD and A has coordinates (1, 4), find the equation of the locus of all points in the plane whose distance from A is $|AM|$.

6. X is the point (−3, 4) and Y is the point (−1, −5). A variable point Z (x, y) is such that $ZX^2 − ZY^2 = 15$. Find the equation of the locus Z.

7. Two points A and B have coordinates (1, 2) and (2, −3) respectively. P is a locus given by:

$|AP|^2 + |BP|^2 = 10$.

What is the equation of the locus P?

8. Find the equation of the locus of points which is equidistant from points S(2, −3) and T(1, 4).

CHAPTER 4
COORDINATE GEOMETRY OF CIRCLE

A circle is the locus of points equidistant from a fixed point. This means that a circle is a path which is at a fixed distance from a given point. A circle can be completely described by its centre and radius. If the coordinates of the centre of a circle of radius r, is given by (a, b) then the equation of the circle is given by:

$$(x-a)^2 + (y-b)^2 = r^2$$

For a circle whose centre is at the origin (0, 0), the equation of the circle is:

$$x^2 + y^2 = r^2 \quad \text{(Since a and b are 0)}$$

The General Equation of a Circle

The general equation of a circle of center (a, b) and radius r, is given by:

$$x^2 + y^2 - 2ax - 2ay + c = 0$$

where: $c = a^2 + b^2 - r^2$

This general equation shows that:
(a) the highest power of x and y is 2
(b) the coefficient of x^2 and y^2 must be equal
(c) there is no term in xy

The Equation of a Circle When Given the Center and a Point

When the center (a, b) and a point, (x_1, y_1) on a circle are given, the equation of the circle is given by:

$$x^2 + y^2 - 2ax - 2by + a^2 + b^2 - (x_1 - a)^2 - (y_1 - b)^2 = 0$$

Examples

1. Find the equation of a circle of radius 4 units and center at the origin.

<u>Solution</u>

Since the center is at the origin, the equation is given by:

$$x^2 + y^2 = r^2$$
$$x^2 + y^2 = 4^2$$
$$x^2 + y^2 = 16$$

2. Determine the equation of the circle of center (−2, 5) and radius 3 units

<u>Solution</u>

Method 1

The equation of a circle is given by:

$$(x-a)^2 + (y-b)^2 = r^2$$

Substitute −2 for a, 5 for b and 3 for r. This gives:

$$(x-(-2))^2 + (y-5)^2 = 3^2$$
$$(x+2)^2 + (y-5)^2 = 9$$
$$x^2 + 4x + 4 + y^2 - 10y + 25 = 9$$

$$x^2 + y^2 + 4x - 10y + 4 + 25 - 9 = 0$$
$$x^2 + y^2 + 4x - 10y + 20 = 0$$

Method 2

The general equation of a circle is given by:
$$x^2 + y^2 - 2ax - 2by + c = 0$$
But, $\quad c = a^2 + b^2 - r^2$

Therefore the equation becomes:
$$x^2 + y^2 - 2ax - 2by + a^2 + b^2 - r^2 = 0$$
Substituting a = −2, b = 5 and r = 3 gives:
$$x^2 + y^2 - (2(-2)x) - (2(5)y) + (-2)^2 + (5)^2 - 3^2 = 0$$
$$x^2 + y^2 + 4x - 10y + 4 + 25 - 9 = 0$$
$$x^2 + y^2 + 4x - 10y + 20 = 0$$

3. The center of a circle is at the point (3, 4) and the circle passes through (−1, 2). Find:
(a) the radius of the circle
(b) the equation of the circle.

Solution

Method 1

The radius of the circle is the distance between the center and the point on the circumference. Taking (a, b) as (3, 4) and (x_1, y_1) as (−1, 2), the radius which is the distance between these two points is given by:
$$r^2 = (x_1 - a)^2 + (y_1 - b)^2$$
$$= (-1 - 3)^2 + (2 - 4)^2$$
$$= (-4)^2 + (-2)^2$$
$$= 16 + 4$$
$$r^2 = 20$$
$$r^2 = \sqrt{20}$$
$$r = 4.5 \text{ units}$$

(b) The equation of the circle is given by:
$$(x - a)^2 + (y - b)^2 = r^2$$
$$(x - 3)^2 + (y - 4)^2 = 20 \quad \text{(Note that } r^2 = 20 \text{ from (a) above)}$$
$$x^2 - 6x + 9 + y^2 - 8y + 16 = 20$$
$$x^2 + y^2 - 6x - 8y + 9 + 16 - 20 = 0$$
$$x^2 + y^2 - 6x - 8y + 5 = 0$$

Method 2

When given the center and a center and a point, the equation of a circle is given by:
$$x^2 + y^2 - 2ax - 2by + a^2 + b^2 - (x_1 - a)^2 - (y_1 - b)^2 = 0$$
Substituting a = 3, b = 4, x_1 = −1, y_1 = 2, we obtain the equation as follows:
$$x^2 + y^2 - (2(3)x) - (2(4)y) + 3^2 + 4^2 - (-1 - 3)^2 - (2 - 4)^2 = 0$$
$$x^2 + y^2 - 6x - 8y + 9 + 16 - (-4)^2 - (-2)^2 = 0$$

$$x^2 + y^2 - 6x - 8y + 9 + 16 - 16 - 4 = 0$$
$$x^2 + y^2 - 6x - 8y + 5 = 0$$

(a) Recall that $c = a^2 + b^2 - r^2$

Hence, $r^2 = a^2 + b^2 - c$

$\qquad = 3^2 + 4^2 - 5 \qquad$ (Note that c = 5 from the equation of the circle above)

$\qquad = 9 + 16 - 5$

$\quad r^2 = 20$

$\quad r = \sqrt{20}$

$\quad r = 4.5$ units

(b) The equation of the circle is:

$$x^2 + y^2 - 6x - 8y + 5 = 0 \qquad \text{(As obtained above)}$$

4. The coordinates of the ends of the diameter of a circle are given by (−3, 7) and (5, −1). Find:
(a) the radius of the circle
(b) the equation of the circle

Solution

The center of the circle is the mid-point of the diameter. This is given by:

$$\text{Midpoint (center)} = \left(\frac{-3+5}{2}, \ \frac{7+(-1)}{2} \right)$$

$\qquad$ Center = (1, 3)

We can now use the center and one of the points at the ends of the diameter to obtain the equation of the circle. Let us use the point (5, −1) as (x_1, y_1) and the center (1, 3) as (a, b). Hence, the equation is given by:

$$x^2 + y^2 - 2ax - 2by + a^2 + b^2 - (x_1 - a)^2 - (y_1 - b)^2 = 0$$
$$x^2 + y^2 - (2(1)x) - (2(3)y) + 1^2 + 3^2 - (5-1)^2 - (-1-3)^2 = 0$$
$$x^2 + y^2 - 2x - 6y + 1 + 9 - (4)^2 - (-4)^2 = 0$$
$$x^2 + y^2 - 2x - 6y + 1 + 9 - 16 - 16 = 0$$
$$x^2 + y^2 - 2x - 6y - 22 = 0$$

(a) The radius of the circle is given by:

$\quad r^2 = a^2 + b^2 - c$

$\qquad = 1^2 + 3^2 - (-22)$

$\qquad = 1 + 9 + 22$

$\quad r^2 = 32$

$\quad r = \sqrt{32}$

$\quad r = 5.7$ units

(b) The equation of the circle is:

$$x^2 + y^2 - 2x - 6y - 22 = 0$$

5. The equation of a circle is given by: $x^2 + y^2 + 8x - 2y + 1 = 0$. Find the:
(a) centre

(b) radius of the circle.

Solution

Arranging the equation above and the general equation of a circle gives:

$$x^2 + y^2 + 8x - 2y + 1 = 0$$

General: $x^2 + y^2 - 2ax - 2by + c = 0$

Comparing the x part of the two equations above shows that:

$$-2a = 8$$
$$a = \frac{8}{-2}$$
$$a = -4$$

Similarly, comparing the y part of the two equations above shows that:

$$-2b = -2$$
$$b = \frac{-2}{-2}$$
$$b = 1$$

Finally, the constant part of the equation above shows that c = 1

(a) Therefore the center of the circle is given by:

Center = (−4, 1)   (i.e. the values of a and b)

(b) The radius of the circle is given by:

$$r^2 = a^2 + b^2 - c$$
$$= (-4)^2 + (1)^2 - 1$$
$$= 16 + 1 - 1$$
$$r^2 = 16$$
$$r = \sqrt{16}$$
$$r = 4$$

Therefore, the radius of the circle is 4 units

6. Determine the center and radius of a circle whose equation is: $5x^2 + 5y^2 - 10x + 15y - 3 = 0$

Solution

$$5x^2 + 5y^2 - 10x + 15y - 3 = 0$$

We first make the coefficient of x^2 and y^2 to be 1 by dividing each term by 5. This gives:

$$x^2 + y^2 - 2x + 3y - \frac{3}{5} = 0$$

General equation: $x^2 + y^2 - 2ax - 2by + c = 0$

Comparing the two equations above that:

$$-2a = -2$$
$$a = \frac{-2}{-2}$$
$$a = 1$$

Also, $-2b = 3$
$$b = -\frac{3}{2}$$

And $c = -\frac{3}{5}$

Therefore the center of the circle is given by:

$$\text{Center} = (1, -\frac{3}{2}) \quad \text{(i.e. the values of a and b)}$$

(b) The radius of the circle is given by:

$$r^2 = a^2 + b^2 - c$$

$$= (1)^2 + (-\frac{3}{2})^2 - (-\frac{3}{5})$$

$$= 1 + \frac{9}{4} + \frac{3}{5}$$

$$= \frac{20 + 45 + 12}{20}$$

$$r^2 = \frac{77}{20}$$

$$r = \sqrt{\frac{77}{20}}$$

$$r = 1.96 \text{ units}$$

Therefore, the radius of the circle is 1.96 units

7. Find the equation of the circle which lies on the points (4, 5), (3 , −2) and (−6 −3).

Solution

The general equation of a circle is:

$$x^2 + y^2 - 2ax - 2by + c = 0$$

Substituting the first point (4, 5) into this equation gives:

$$4^2 + 5^2 - 2a(4) - 2b(5) + c = 0$$

$$16 + 25 - 8a - 10b + c = 0$$

$$8a + 10b - c = 41 \text{ ....(1)}$$

Substituting the second point (3, −2) into the equation gives:

$$3^2 + (-2)^2 - 2a(3) - 2b(-2) + c = 0$$

$$9 + 4 - 6a + 4b + c = 0$$

$$6a - 4b - c = 13 \text{ ....(2)}$$

Substituting the third point (−6, −3) into the equation gives:

$$(-6)^2 + (-3)^2 - 2a(-6) - 2b(-3) + c = 0$$

$$36 + 9 + 12a + 6b + c = 0$$

$$12a + 6b + c = -45 \text{ ....(3)}$$

Bringing equations (1), and (2) and (3) together gives:

$$8a + 10b - c = 41 \text{ ....(1)}$$

$$6a - 4b - c = 13 \text{ ....(2)}$$

$$12a + 6b + c = -45 \text{ ....(3)}$$

Let us eliminate c as follows:

Equations (1) – (2) gives: 2a + 14b = 28. Divide each term by 2 to obtain:

$$a + 7b = 14 \text{ ....(4)}$$

53

Equation (1) + (3) gives:

$20a + 16b = -4$

Or,　$5a + 4b = -1$ ....(5)　　(After dividing each term by 4)

From equation (4):　$a = 14 - 7b$ ....(6)

Substitute $14 - 7b$ for a in equation (5). This gives:

$5a + 4b = -1$ ....(5)

$5(14 - 7b) + 4b = -1$

$70 - 35b + 4b = -1$

$70 + 1 = 35b - 4b$

$31b = 71$

$b = \dfrac{71}{31}$

From equation (6) we obtain a as follows:

$a = 14 - 7b$ ....(6)

$= 14 - 7(\dfrac{71}{31})$　　(Since $b = \dfrac{71}{31}$)

$= \dfrac{434 - 497}{31}$

$a = -\dfrac{63}{31}$

Let us substitute the values of a and b into equation (2) in order to obtain the value of c

$6a - 4b - c = 13$ ....(2)

$6(-\dfrac{63}{31}) - 4(-\dfrac{71}{31}) - c = 13$ ....(2)

$-\dfrac{378}{31} - \dfrac{284}{31} - 13 = c$

$\dfrac{-378 - 284 - 403}{31} = c$

$c = -\dfrac{1065}{31}$

Therefore the equation of the circle is given by:

$x^2 + y^2 - 2ax - 2by + c = 0$

$x^2 + y^2 - 2(-\dfrac{63}{31})x - 2(\dfrac{71}{31})y + (-\dfrac{1065}{31}) = 0$

$x^2 + y^2 + \dfrac{126}{31}x - \dfrac{142}{31}y - \dfrac{1065}{31} = 0$

Multiply each term by 31 in order to clear out the fractions. This gives:

$31x^2 + 31y^2 + 126x - 142y - 1065 = 0$

8. A circle passes through the points (−3, −4), (−6, 5) and (2, 1). Find:

(a) the radius of the circle

(b) the equation of the circle

Solution

Method 1

The general equation of a circle is:

$$x^2 + y^2 - 2ax - 2by + c = 0$$

Substituting the first point (−3, −4) into this equation gives:

$$(-3)^2 + (-4)^2 - 2a(-3) - 2b(-4) + c = 0$$

$$9 + 16 + 6a + 8b + c = 0$$

$$6a + 8b + c = -25 \ldots\ldots\ldots\ldots(1)$$

Substituting the second point (−6, 5) into the equation gives:

$$(-6)^2 + (5)^2 - 2a(-6) - 2b(5) + c = 0$$

$$36 + 25 + 12a - 10b + c = 0$$

$$12a - 10b + c = -61 \ldots\ldots\ldots\ldots(2)$$

Substituting the third point (2, 1) into the equation gives:

$$(2)^2 + (1)^2 - 2a(2) - 2b(1) + c = 0$$

$$4 + 1 - 4a - 2b + c = 0$$

$$4a + 2b - c = 5 \ldots\ldots\ldots\ldots(3)$$

Bringing equations (1), and (2) and (3) together gives:

$$6a + 8b + c = -25 \ldots\ldots\ldots\ldots(1)$$

$$12a - 10b + c = -61 \ldots\ldots\ldots\ldots(2)$$

$$4a + 2b - c = 5 \ldots\ldots\ldots\ldots(3)$$

Let us eliminate c as follows:

Equations (2) − (1) gives: $6a - 18b = -36$. Divide each term by 6 to obtain:

$$a - 3b = -6 \ldots\ldots\ldots\ldots\ldots(4)$$

Equation (1) + (3) gives: $10a + 10b = -20$. Divide each term by 10 to obtain:

$$a + b = -2 \ldots\ldots\ldots\ldots\ldots(5)$$

Bringing equation (4) and (5) together to solve them simultaneously gives:

$$a - 3b = -6 \ldots\ldots\ldots\ldots\ldots(4)$$

$$\underline{a + b = -2} \ldots\ldots\ldots\ldots\ldots(5)$$

Equation (5) − (4): $4b = 4$

$$b = \frac{4}{4}$$

$$b = 1$$

Put b = 1 in equation (5) in order to find a.

$$a + b = -2 \ldots\ldots\ldots\ldots\ldots(5)$$

$$a + 1 = -2 \ldots\ldots\ldots\ldots\ldots(5)$$

$$a = -2 - 1$$

$$a = -3$$

Let us substitute the values of a and b into equation (3) in order to obtain the value of c

$4a + 2b - c = 5$ ....(3)

$4(-3) + 2(1) - c = 5$

$-12 + 2 - c = 5$

$c = -12 + 2 - 5$

$c = -15$

Therefore the equation of the circle is given by:

$x^2 + y^2 - 2ax - 2by + c = 0$

$x^2 + y^2 - 2(-3)x - 2(1)y + (-15) = 0$

$x^2 + y^2 + 6x - 2y - 15 = 0$

Method 2

Let the points be A = (−3, −4), B = (−6, 5) and C = (2, 1). Let the center of the circle be O.

Therefore, the distance from O to any of the points A, B and C gives the radius of the circle.

Hence, $|OA|^2 = |OB|^2 = |OC|^2$

where radius, r = $|OA|$, $|OB|$ or $|OC|$

With coordinates of the centre, O, as (a, b) we proceed as follows:

$|OA|^2 = |OB|^2 = |OC|^2$

$(-3 - a)^2 + (-4 - b)^2 = (-6 - a)2 + (5 - b)^2 = (2 - a)^2 + (1 - b)^2$

$9 + 6a + a^2 + 16 + 8b + b^2 = 36 + 12a + a^2 + 25 - 10b + b^2 = 4 - 4a + a^2 + 1 - 2b + b^2$

$a^2 + b^2 + 6a + 8b + 25 = a^2 + b^2 + 12a - 10b + 61 = a^2 + b^2 - 4a - 2b + 5$

We take the first and second parts of the equation and simplify as follows:

$a^2 + b^2 - a^2 - b^2 + 6a - 12a + 8b + 10b + 25 - 61 = 0$

$-6a + 18b = 36$

$-a + 3b = 6$ (After dividing by 6)

Or, $a - 3b = -6$ ....(1)

Similarly, we take the second and third parts of the equation and simplify as follows:

$a^2 + b^2 + 12a - 10b + 61 = a^2 + b^2 - 4a - 2b + 5$

$a^2 + b^2 - a^2 - b^2 + 12a + 4a - 10b + 2b + 61 - 5 = 0$

$16a - 8b = -56$

Or, $2a - b = -7$ ....(2)

We now solve equation (1) and (2) simultaneously as follows:

$a - 3b = -6$ ....(1)

$2a - b = -7$ ....(2)

From equation (1), $a = 3b - 6$ ....(3)

Substitute 3b − 6 for a in equation (2). This gives:

$2a - b = -7$ ....(2)

$2(3b - 6) - b = -7$

$6b - 12 - b = -7$

$5b = -7 + 12$

$b = \dfrac{5}{5}$

$b = 1$

Put b = 1 in equation (3) to find a as follows:

$a = 3b - 6$ ....(3)

$= 3(1) - 6$

$= 3 - 6$

$a = -3$

Therefore, the center of the circle, (a, b) is (−3, 1)

(a) The radius of the circle is the distance from the center (−3, 1) to any point on the circle. Let us use the point (2, 1). Therefore:

$r^2 = (x_1 - a)^2 + (y_1 - b)^2$

$= (2 - (-3)^2 + (1 - 1)^2$

$= (2 + 3)^2 + 0^2$

$r^2 = 25$

$r = 5$

(b) The equation of the circle is obtained as follows:

$(x - a)^2 + (y - b)^2 = r^2$

$(x - (-3))^2 + (y - 1)^2 = 25$ (Note that $r^2 = 25$ from (a) above)

$(x + 3)^2 + (y - 1)^2 = 25$

$x^2 + 6x + 9 + y^2 - 2y + 1 = 25$

$x^2 + y^2 + 6x - 2y + 10 = 25$

$x^2 + y^2 + 6x - 2y - 15 = 0$

Equation of Tangent and Normal to a Circle

A tangent is an external line that touches the circumference of a circle. A line that passes the center of a circle, and passes through the point where the tangent meets the circle is called a normal. The normal and the tangent are perpendicular to each other.

The equation of a tangent to a circle at the point (x_1, y_1) on the circle is given by:

$xx_1 + yy_1 - a(x + x_1) - b(y + y_1) + c = 0$

The equation of a normal to a circle at the point (x_1, y_1) on the circle is given by:

$xy_1 - yx_1 + a(y - y_1) - b((x - x_1) = 0$

Length of a Tangent to a Circle from an External Point

The length of a tangent to a circle from an external point (x_1, y_1) is given by:

$L^2 = (a - x_1)^2 + (b - y_1)^2 - r^2$

Examples

1. The equation of a circle is $x^2 + y^2 - 6x + 2y - 42 = 0$. Determine the equation of the tangent to the circle at the point $(-1, 5)$ on the circle.

Solution

Method 1

Let us first find the center (a, b) of the circle.
$$x^2 + y^2 - 6x + 2y - 42 = 0$$

General equation: $x^2 + y^2 - 2ax - 2by + c = 0$

Comparing terms shows that

$$-2a = -6$$
$$a = \frac{-6}{-2}$$
$$a = 3$$

Also, $-2b = 2$
$$b = \frac{2}{-2}$$
$$b = -1$$

Hence, the center, (a, b) of the circle is $(3, -1)$. From the equation of the circle, $c = -42$. Note that (x_1, y_1) is $(-1, 5)$ as given in the question. Therefore the equation of the tangent to the circle is given by:

$$xx_1 + yy_1 - a(x + x_1) - b(y + y_1) + c = 0$$
$$-1x + 5y - 3(x - 1) - (-1)(y + 5) - 42 = 0$$
$$-x + 5y - 3x + 3 + y + 5 - 42 = 0$$
$$-4x + 6y - 34 = 0$$
$$6y - 4x - 34 = 0$$

Or, $3y - 2x - 17 = 0$ (After dividing each term by 2)

Method 2

After finding the center (a, b) i.e. $(3, -1)$, we use it to find the gradient of the normal which also passes (x_1, y_1) i.e. $(-1, 5)$. Hence, by using these two points we first find the gradient of the normal as follows:

$$m_1 = \frac{y_2 - y_1}{x_2 - x_1}$$
$$= \frac{5 - (-1)}{-1 - 3}$$
$$= \frac{6}{-4}$$
$$m_1 = -\frac{3}{2}$$

Since the normal and the tangent are perpendicular, then the gradient of the tangent is $m_2 = \frac{2}{3}$ (Recall that $m_1 m_2 = -1$). Hence, the equation of the tangent with slope $\frac{2}{3}$ and passing through $(-1, 5)$ is given by:

$$m_2 = \frac{y - y_1}{x - x_1}$$

$$\frac{2}{3} = \frac{y-5}{x-(-1)}$$

$2(x+1) = 3(y-5)$

$2x + 2 = 3y - 15$

$2x - 3y + 17 = 0$

Or, $3y - 2x - 17 = 0$

2. Find the equation of the tangent to the circle $x^2 + y^2 - 10x + 8y - 59 = 0$ at the point (−3, 2).

<u>Solution</u>

Let us first find the center (a, b) of the circle.

$$x^2 + y^2 - 10x + 8y - 59 = 0$$

General equation: $x^2 + y^2 - 2ax - 2by + c = 0$

Comparing terms shows that

$-2a = -10$

$a = \dfrac{-10}{-2}$

$a = 5$

Also, $-2b = 8$

$b = \dfrac{8}{-2}$

$b = -4$

Hence, the center, (a, b) of the circle is (5, −4). From the equation of the circle, c = −59. Note that (x_1, y_1) is (−3, 2) as given in the question. Therefore the equation of the tangent to the circle is given by:

$xx_1 + yy_1 - a(x + x_1) - b(y + y_1) + c = 0$

$-3x + 2y - 5(x - 3) - (-4)(y + 2) - 59 = 0$

$-3x + 2y - 5x + 15 + 4y + 8 - 59 = 0$

$-8x + 6y - 36 = 0$

$6y - 8x - 36 = 0$

Or, $3y - 4x - 18 = 0$ (After dividing each term by 2)

Hence, the equation of the tangent to the circle is given by $3y - 4x - 18 = 0$

3. The equation of a circle is $x^2 + y^2 + 3x + 5y - 18 = 0$. Determine the equation of the normal to the circle at the point (1, 2) on the circle.

<u>Solution</u>

Method 1

Let us first find the center (a, b) of the circle.

$$x^2 + y^2 + 3x + 5y - 18 = 0$$

General equation: $x^2 + y^2 - 2ax - 2by + c = 0$

Comparing terms shows that

$-2a = 3$

$a = \dfrac{3}{-2}$

$$a = -\frac{3}{2}$$

Similarly, $-2b = 5$

$$b = \frac{5}{-2}$$

$$b = -\frac{5}{2}$$

Hence, the center, (a, b) of the circle is $(-\frac{3}{2}, -\frac{5}{2})$. Note that (x_1, y_1) is (1, 2) as given in the question. Therefore the equation of the normal to the circle is given by:

$$xy_1 - yx_1 + a(y - y_1) - b(x - x_1) = 0$$
$$2x - 1y - \frac{3}{2}(y - 2) - (-\frac{5}{2})(x - 1) = 0$$
$$2x - y - \frac{3}{2}y + 3 + \frac{5}{2}x - \frac{5}{2} = 0$$

Multiplying each term by 2 to clear out the fractions gives:

$$4x - 2y - 3y + 6 + 5x - 5 = 0$$
$$9x - 5y + 1 = 0$$

Method 2

After finding the points a and b, then the equation of the normal is obtained by finding the equation of the line through the points (a, b) and (x_1, y_1). Note that the normal is a radius /diameter line which must through the center (a, b) of the circle. Hence, the normal is given by the equation of a straight line through two points as follows:

$$\frac{y - y_1}{x - x_1} = \frac{y_2 - y_1}{x_2 - x_1}$$

$$\frac{y - 2}{x - 1} = \frac{-\frac{5}{2} - 2}{-\frac{3}{2} - 1}$$

$$\frac{y - 2}{x - 1} = \frac{-\frac{9}{2}}{-\frac{5}{2}}$$

$$\frac{y - 2}{x - 1} = -\frac{9}{2}x - \frac{2}{5}$$

$$\frac{y - 2}{x - 1} = \frac{9}{5}$$

$$9(x - 1) = 5(y - 2)$$
$$9x - 9 = 5y - 10$$
$$9x - 5y + 1 = 0 \quad \text{(As obtained in method 1)}$$

Note that this equation can also be written as:

$$5y - 9x - 1 = 0 \quad \text{(When all terms are moved to the other side of the equation)}$$

4. Find the equation of the normal to the circle $3x^2 + 3y^2 - 9x - 2y - 17 = 0$ at the point (−1, −1).

<u>Solution</u>

$$3x^2 + 3y^2 - 9x - 2y - 17 = 0$$

Dividing each term by 3 gives:

$$x^2 + y^2 - 3x - \frac{2}{3}y - \frac{17}{3} = 0$$

General equation: $x^2 + y^2 - 2ax - 2by + c = 0$

Comparing terms shows that

$$-2a = -3$$

$$a = \frac{-3}{-2}$$

$$a = \frac{3}{2}$$

Similarly, $-2b = -\frac{2}{3}$

$$b = -\frac{2}{3}(-\frac{1}{2})$$

$$b = \frac{1}{3}$$

Hence, the center, (a, b) of the circle is $(\frac{3}{2}, \frac{1}{3})$, while (x_1, y_1) is $(-1, -1)$. Therefore the equation of the normal to the circle is given by:

$$xy_1 - yx_1 + a(y - y_1) - b(x - x_1) = 0$$

$$-1x - (-1)y + \frac{3}{2}(y - (-1)) - (\frac{1}{3})(x - (-1)) = 0$$

$$-x + y + \frac{3}{2}y + \frac{3}{2} + \frac{1}{3}x - \frac{1}{3} = 0$$

Multiplying each term by 6 (i.e. the LCM of 2 and 3) to clear out the fractions gives:

$$-6x + 6y + 9y + 9 + 2x - 2 = 0$$

$$-4x + 15y + 7 = 0$$

$$15y - 4x + 7 = 0$$

5. Find the length of the tangent to the circle $x^2 + y^2 - 3x + 2y - 10 = 0$ from the point $(-5, 2)$

<u>Solution</u>

$$x^2 + y^2 - 3x + 2y - 10 = 0$$

General equation: $x^2 + y^2 - 2ax - 2by + c = 0$

Comparing terms shows that

$$-2a = -3$$

$$a = \frac{-3}{-2}$$

$$a = \frac{3}{2}$$

Similarly, $-2b = 2$

$$b = \frac{2}{-2}$$

$$b = -1$$

Hence, the center, (a, b) of the circle is $(\frac{3}{2}, -1)$

The radius of the circle is given by:

$$r^2 = a^2 + b^2 - c$$

$$= (\frac{3}{2})^2 + (-1)^2 - (-10) \quad \text{(Note that c is the constant term in the equation of the circle)}$$

$$= \frac{9}{4} + 1 + 10$$

$$= \frac{9 + 4 + 40}{4}$$

$$r^2 = \frac{53}{4}$$

The length of the tangent is given by:

$$L^2 = (a - x_1)^2 + (b - y_1)^2 - r^2$$

$$L^2 = (\frac{3}{2} - (-5))^2 + (-1 - 2)^2 - \frac{53}{4} \quad \text{(Note that } (x_1, y_1) \text{ is } (-5, 2) \text{ as given in the question)}$$

$$= (\frac{3}{2} + 5)^2 + (-3)^2 - \frac{53}{4}$$

$$= (\frac{13}{2})^2 + 9 - \frac{53}{4}$$

$$= \frac{169}{4} + 9 - \frac{53}{4}$$

$$= \frac{169 + 36 - 53}{4}$$

$$= \frac{152}{4}$$

$$L^2 = 38$$

$$L = \sqrt{38}$$

$$L = 6.16 \text{ units}$$

6. Find the length of the tangent to the circle $2x^2 + 2y^2 + x - 2y - 17 = 0$ from the point (5, −3).

Solution

$$2x^2 + 2y^2 + x - 2y - 17 = 0$$

Dividing each term by 2 in order to make the coefficient of x^2 and y^2 to be 1 will give:

$$x^2 + y^2 + \frac{1}{2}x - y - \frac{17}{2} = 0$$

General equation: $x^2 + y^2 - 2ax - 2by + c = 0$

Comparing terms shows that

$$-2a = \frac{1}{2}$$

$$a = \frac{1}{2}(\frac{1}{-2})$$

$$a = -\frac{1}{4}$$

Similarly, $-2b = -1$

$$b = \frac{-1}{-2}$$

$$b = \frac{1}{2}$$

Hence, the center, (a, b) of the circle is $(-\frac{1}{4}, \frac{1}{2})$

The radius of the circle is given by:

$$r^2 = a^2 + b^2 - c$$

$$= (-\frac{1}{4})^2 + (\frac{1}{2})^2 - (-17)$$

$$= \frac{1}{16} + \frac{1}{4} + 17$$

$$= \frac{1 + 4 + 272}{16}$$

$$r^2 = \frac{277}{16}$$

The length of the tangent is given by:

$$L^2 = (a - x_1)^2 + (b - y_1)^2 - r^2$$

$$L^2 = (-\frac{1}{4} - 5)^2 + (\frac{1}{2} - (-3))^2 - \frac{277}{16} \qquad \text{(Note that } (x_1, y_1) \text{ is (5, –3) as given in the question)}$$

$$= (-\frac{21}{4})^2 + (\frac{7}{2})^2 - \frac{277}{16}$$

$$= \frac{441}{16} + \frac{49}{4} - \frac{277}{16}$$

$$= \frac{441 + 196 - 277}{16}$$

$$= \frac{360}{16}$$

$$L^2 = \frac{45}{2}$$

$$L = \sqrt{\frac{45}{2}}$$

$$L = 4.74 \text{ units}$$

Miscellaneous Problems on Circle and Tangent to a Circle

1. Which of the following equations describe a circle? If not give reasons for your answers.

(a) $x^2 + 2y^2 = 8$

(b) $2x^2 + 2y^2 + 8x - 7y = 0$

(c) $x^2 + y^2 - 5x + 10y = 0$

(d) $2x^2 - 2y^2 + 3x + 11y = 0$

(e) $7x^2 + 7y^2 - 12x = 18$

(f) $x^2 + y^2 - 5xy - 12 = 0$

(g) $x^2 + y^3 - 2x + 5y - 9 = 0$

(h) $3x^2 + 3y^2 - 8x + 17y = 0$

(i) $20x^2 + 20y^2 - 7x + 15y + 12 = 0$

Solution

(a) It is not a circle. The coefficient of x^2 and y^2 are not the same.

(b) It is a circle

(c) It is a circle

(d) It is not a circle. The coefficient of x^2 and y^2 are not the same. They are 2 for x^2 and −2 for y^2.

(e) It is a circle

(f) It is not a circle. A circle does not contain xy.

(g) It is not a circle. The greatest power of x and y is not 2. That of y is 3

(h) It is a circle

(i) It is a circle

2. Show that the point (2, 3) lies on the circle: $x^2 + y^2 - 3x + 5y - 22 = 0$

Solution

Substitute the point (2, 3) into the left hand side of the given equation.

LHS: $x^2 + y^2 - 3x + 5y - 22$

$\quad 2^2 + 3^2 - 3(2) + 5(2) - 22$

$\quad 4 + 9 - 6 + 10$

$\quad = 0$

Since the left hand side (LHS) is equal to zero, just like the right hand side (RHS), it shows that the point (2, 3) lies on the circle.

3. The line $x + 7y - 5 = 0$ cuts the circle $x^2 + y^2 = 25$ at A and B. Find the points A and B.

Solution

We have to solve the two equations simultaneously.

$\quad x + 7y - 5 = 0$ ....(1)

$\quad x^2 + y^2 = 25$ ....(2)

From equation (1), $x = 5 - 7y$ ....(3)

Put $x = 5 - 7y$ in equation (2) as follows:

$\quad x^2 + y^2 = 25$ ....(2)

$\quad (5 - 7y)^2 + y^2 = 25$

$\quad 25 - 70y + 49y^2 + y^2 = 25$

$\quad 50y2 - 70y = 0$

$\quad 5y^2 - 7y = 0$ (After dividing by 10)

$\quad y(5y - 7) = 0$

Hence, y = 0

and $5y - 7 = 0$

$$y = \frac{7}{5}$$

Therefore, y = 0 and y = $\frac{7}{5}$

When y = 0 we find x from equation (3) as follows:

$$x = 5 - 7y \ldots\ldots\ldots\ldots(3)$$
$$= 5 - 7(0)$$
$$x = 5$$

When y = $\frac{7}{5}$ we obtain x as follows

$$x = 5 - 7y \ldots\ldots\ldots\ldots(3)$$
$$= 5 - 7\left(\frac{7}{5}\right)$$
$$= 5 - \frac{49}{5}$$
$$= \frac{25 - 49}{5}$$
$$x = \frac{-24}{5}$$

When x = 5, y = 0, so the point A is (5, 0). When $x = -\frac{24}{5}$, y = $\frac{7}{5}$, so the point B is $\left(-\frac{24}{5}, \frac{7}{5}\right)$

Therefore the points A and B are (5, 0) and $\left(-\frac{24}{5}, \frac{7}{5}\right)$

4. M(2, 1) and N(4, 5) are two points on an x-y plane. Point P(x, y) moves in the plane such that |MP| : |NP| = 2 : 1
(a) Determine the equation of the locus P
(b) Hence, describe the locus completely
Solution
(a) |MP| : |NP| = 2 : 1
This also means that:

$$\frac{|MP|}{|NP|} = \frac{2}{1}$$
$$\frac{|MP|^2}{|NP|^2} = \frac{2^2}{1^2}$$

With MP as distance between M and P, and NP as distance between N and P we substitute the corresponding coordinates of the points as follows:

$$\frac{(x-2)^2 + (y-1)^2}{(x-4)^2 + (y-5)^2} = \frac{2^2}{1^2}$$
$$\frac{x^2 - 4x + 4 + y^2 - 2y + 1}{x^2 - 8x + 16 + y^2 - 10y + 25} = \frac{4}{1}$$

$$\frac{x^2 + y^2 - 4x - 2y + 5}{x^2 + y^2 - 8x - 10y + 41} = \frac{4}{1}$$

$$4(x^2 + y^2 - 8x - 10y + 41) = x^2 + y^2 - 4x - 2y + 5$$
$$4x^2 + 4y^2 - 32x - 40y + 164 = x^2 + y^2 - 4x - 2y + 5$$
$$4x^2 - x^2 + 4y^2 - y^2 - 32x + 4x - 40y + 2y + 164 - 5 = 0$$
$$3x^2 + 3y^2 - 28x - 38y + 159 = 0$$

(b) The locus is a circle. Let us find the center and radius of the circle.
$$3x^2 + 3y^2 - 28x - 38y + 159 = 0$$

Dividing each term by 3 to make the coefficient of x^2 and y^2 to be 1 gives:
$$x^2 + y^2 - \frac{28}{3}x - \frac{38}{3}y + 53 = 0$$

General equation: $x^2 + y^2 - 2ax - 2by + c = 0$

Comparing corresponding terms shows that:
$$-2a = -\frac{28}{3}$$

$$a = -\frac{28}{3} \div -2$$

$$= -\frac{28}{3}\left(-\frac{1}{2}\right)$$

$$a = \frac{14}{3}$$

Also, $-2b = -\dfrac{38}{3}$

$$b = -\frac{38}{3} \div -2$$

$$= -\frac{38}{3}\left(-\frac{1}{2}\right)$$

$$b = \frac{19}{3}$$

The radius is obtained as follows:
$$r^2 = a^2 + b^2 - c$$

$$= \left(\frac{14}{3}\right)^2 + \left(\frac{19}{3}\right)^2 - 53$$

$$= \frac{196}{9} + \frac{361}{9} - 53$$

$$= \frac{196 - 361 - 477}{9}$$

$$r^2 = \frac{80}{9}$$

$$r = \sqrt{\frac{80}{9}}$$

$$r = \frac{\sqrt{80}}{3}$$

Therefore, the locus is a circle of radius $\frac{\sqrt{80}}{3}$ and center $\left(\frac{14}{3}, \frac{19}{3}\right)$

5. A point E moves in an x-y plane such that its distance from the origin O is twice its distance from the point (3, 0)

(a) Show that the locus of E is a circle

(b) find its center and radius

Solution

The coordinates of O is (0, 0), i.e. the origin.

Let the coordinates of E be (x, y)

The distance from E to O is $|OE|$

The distance from E to P is $|PE|$

Therefore, from the question:

$\quad |OE| = 2|PE|$

Or $\quad |OE|^2 = 2^2|PE|^2$ (When both sides are squared)

Hence, $(x-0)^2 + (y-0)^2 = 4[(x-3)^2 + (y-0)^2]$

$\quad x^2 + y^2 = 4(x^2 - 6x + 9 + y^2)$

$\quad x^2 + y^2 = 4x^2 - 24x + 36 + 4y^2$

$\quad 0 = 4x^2 - x^2 + 4y^2 - y^2 - 24x + 36$

$\quad 3x^2 + 3y^2 - 24x + 36 = 0$

or, $\quad x^2 + y^2 - 8x + 12 = 0$ (After dividing each term by 3)

Since this equation is that of a circle, it shows that the locus of E is a circle.

(b) $\qquad\qquad x^2 + y^2 - 8x + 12 = 0$

General equation: $x^2 + y^2 - 2ax - 2by + c = 0$

Comparing terms shows that:

$\quad -2a = -8$

$\quad a = \frac{-8}{-2}$

$\quad a = 4$

Also, $\quad -2b = 0$ (Since there is no term in y we use 0)

$\quad b = \frac{0}{-2}$

$\quad b = 0$

The radius is obtained as follows:

$\quad r^2 = a^2 + b^2 - c$

$\qquad = 4^2 + 0^2 - 12$

$\qquad = 16 - 12$

$\quad r^2 = 4$

$\quad r = \sqrt{4}$

$\quad r = 2$

Therefore the center of the circle is (4, 0), i.e. (a, b), while the radius of the circle is 2 units.

6. Find the equation of a circle whose center is (4, −1) and circumference 6π.

Solution

The circumference of a circle is given by $2\pi r$.

Hence, $2\pi r = 6\pi$

$$r = \frac{6\pi}{2\pi}$$

$$r = 3$$

Therefore, with center (4, −1) and radius 3, the equation of the circle is obtained as follows:

$$(x - a)^2 + (y - b)^2 = r^2$$
$$(x - 4)^2 + (y - (-1))^2 = 3^2$$
$$(x - 4)^2 + (y + 1)^2 = 3^2$$
$$x^2 - 8x + 16 + y^2 + 2y + 1 = 9$$
$$x^2 + y^2 - 8x + 2y + 17 - 9 = 0$$
$$x^2 + y^2 - 8x + 2y + 8 = 0$$

7. The equation of a circle is given by:

$$x^2 + y^2 - 10x + 6y + k = 0$$

If the radius of the circle is 9 units, find the value of the constant k.

Solution

$$x^2 + y^2 - 10x + 6y + k = 0$$

General equation: $x^2 + y^2 - 2ax - 2by + c = 0$

Comparing terms shows that:

$$-2a = -10$$
$$a = \frac{-10}{-2}$$
$$a = 5$$

Also, $-2b = 6$

$$b = \frac{6}{-2}$$
$$b = -3$$

Hence, the center of the circle is (5, −3)

The value of k is obtained as follows:

Recall that: $c = a^2 + b^2 - r^2$

Hence, $k = 5^2 + (-3)^2 - 9^2$

$$k = 25 + 9 - 81$$
$$k = -47$$

8. The equation of a circle is given by $2x^2 + 2y^2 - 5x - 8y + 6 = 0$. Find the:

(a) area of the circle

(b) length of the diameter of the circle

Solution

$$2x^2 + 2y^2 - 5x - 8y + 6 = 0$$

Divide each term by 2 to make the coefficient of x^2 and y^2 to be 1. This gives:

$$x^2 + y^2 - \frac{5}{2}x - 4y + 3 = 0$$

General equation: $x^2 + y^2 - 2ax - 2by + c = 0$

Comparing terms shows that:

$$-2a = -\frac{5}{2}$$

$$a = -\frac{5}{2} \div -2$$

$$a = -\frac{5}{2}\left(-\frac{1}{2}\right)$$

$$a = \frac{5}{4}$$

Also, $-2b = -4$

$$b = \frac{-4}{-2}$$

$$b = 2$$

The radius is obtained as follows:

$$r^2 = a^2 + b^2 - c$$

$$= \left(\frac{5}{4}\right)^2 + 2^2 - 3$$

$$= \frac{25}{16} + 4 - 3$$

$$= \frac{25 + 64 - 48}{16}$$

$$r^2 = \frac{41}{16}$$

$$r = \sqrt{\frac{41}{16}}$$

$$r = \frac{\sqrt{41}}{4}$$

The area of the circle is given by:

$$\text{Area} = 2\pi r^2$$

$$= 2\pi\left(\frac{41}{16}\right) \quad \text{(Note that } r^2 = \frac{41}{16} \text{ as shown above)}$$

$$= \frac{41\pi}{8} \text{ square units}$$

(b) Diameter = 2r

$$= 2\left(\frac{\sqrt{41}}{4}\right)$$

$$= \frac{\sqrt{41}}{2} \text{ units}$$

9. The equation of a circle is $3x^2 + 3y^2 + 6x - 3y - 1 = 0$. Find the equation of a straight line which passes through the center of the circle and is parallel to the line $2x + 4y = 7$

Solution

$$3x^2 + 3y^2 + 6x - 3y - 1 = 0$$

Dividing each term by 3 gives:

$$x^2 + y^2 + 2x - y - \frac{1}{3} = 0$$

General equation: $x^2 + y^2 - 2ax - 2by + c = 0$

Comparing terms shows that:

$$-2a = 2$$
$$a = \frac{2}{-2}$$
$$a = -1$$

Also, $-2b = -1$

$$b = \frac{-1}{-2}$$
$$b = \frac{1}{2}$$

Hence, the center, (a, b) of the circle is $(-1, \frac{1}{2})$

Let us find the slope of the straight line as follows:

$$2x + 4y = 7$$
$$4y = -2x + 7$$
$$y = -\frac{1}{2}x + \frac{7}{4} \qquad \text{(After dividing each term by 4)}$$

Hence, the slope of the line is $-\frac{1}{2}$. This means that the slope of the line passing through the center of the circle also has a slope of $-\frac{1}{2}$ since they are parallel lines.

Therefore, the equation of this line passing through the center $(-1, \frac{1}{2})$ and having a slope $-\frac{1}{2}$ is given by:

$$m = \frac{y - y_1}{x - x_1}$$
$$-\frac{1}{2} = \frac{y - \frac{1}{2}}{x - (-1)}$$
$$2(y - \frac{1}{2}) = -1(x + 1)$$
$$2y - 1 = -x - 1$$
$$2y + x = 0$$

10. The line $2x + y + 1 = 0$ passes through the center of a circle. If the circle touched the x-axis at the point (−3, 0), find the equation of the circle.

Solution

When a circle touches the x-axis, it means that the x coordinate of the point where it touches the x-axis (i.e. −3) is also the x coordinate of the center, (a, b), of the circle (i.e. a = −3). This is because the radius of the circle is a line perpendicular to the x-axis. Therefore, the center of the circle is (a, b), which is (−3, b) since a = −3.

The line $2x + y + 1 = 0$ also passes through the center of the circle. Hence, we can find b by substituting (−3, b) into the equation of the line as follows:

$$2x + y + 1 = 0$$
$$2(-3) + b + 1 = 0 \quad \text{(Note that } x = -3 \text{ and } y = b \text{ from } (-3, b), \text{ which is the center of the circle)}$$
$$-6 + b + 1 = 0$$

$$b = 6 - 1$$
$$b = 5$$

Therefore, the center, (a, b) of the circle is (−3, 5)

The radius of the circle is the distance between the center (−3, 5) and the point (−3, 0) on the circle. Since the two points have the same x coordinates, then the distance between them is simply obtained by subtracting the lower value of the y coordinate of the two points from the higher value of the y coordinates of the two points. This is obtained as follows:

$$r = 5 - 0$$
$$r = 5$$

This means that when a circle touches the x-axis, the radius of the circle is b, which is the y coordinate of the center of the circle.

Hence, with center (−3, 5) and radius 5, the equation of the circle is obtained as follows:

$$(x - a)^2 + (y - b)^2 = r^2$$
$$(x - (-3))^2 + (y - 5)^2 = 5^2$$
$$(x + 3)^2 + (y - 5)^2 = 25$$
$$x^2 + 6x + 9 + y^2 - 10y + 25 = 25$$
$$x^2 + y^2 + 6x - 10y + 9 + 25 - 25 = 0$$
$$x^2 + y^2 + 6x - 10y + 9 = 0$$

11. The line $3x - 5y - 1 = 0$ passes through the center of a circle. If the circle touched the x-axis at the point (2, 0), find the equation of the circle.

Solution

By using the information in example 1 as a guide, a = 2.

We find b by substituting (2, b) into the equation of the line as follows:

$$3x - 5y - 1 = 0$$
$$3(2) - 5(b) - 1 = 0$$
$$6 - 5b - 1 = 0$$
$$6 - 1 = 5b$$
$$b = \frac{5}{5}$$
$$b = 1$$

Therefore, the center, (a, b) of the circle is (2, 1)

The radius of the circle is given by:

The radius of the circle is 1, which is the y coordinate of the center of the circle.

Hence, with center (2, 1) and radius 1, the equation of the circle is obtained as follows:

$$(x - a)^2 + (y - b)^2 = r^2$$
$$(x - 2)^2 + (y - 1)^2 = 1^2$$
$$x^2 - 4x + 4 + y^2 - 2y + 1 = 1$$
$$x^2 + y^2 - 4x - 2y + 4 + 1 - 1 = 0$$
$$x^2 + y^2 - 4x - 2y + 4 = 0$$

12. A circle touches the y-axis at the point (0, 4). If the line $x - 5y + 13 = 0$ passes through the center of the circle, find the equation of the circle.

When a circle touches the y-axis, it means that the y coordinate of the point where it touches the y-axis (i.e. 4) is also the y coordinate of the center, (a, b), of the circle (i.e. b = 4). This is because the radius of the circle is a line perpendicular to the y-axis. Therefore, the center of the circle is (a, b), which is (a, 4) since b = 4.

The line $x - 5y + 13 = 0$ also passes through the center of the circle. Hence, we can find a by substituting (a, 4) into the equation of the line as follows:

$x - 5y + 13 = 0$

$a - 5(4) + 13 = 0$

$a - 20 + 13 = 0$

$a = 20 - 13$

$a = 7$

Therefore, the center, (a, b) of the circle is (7, 4)

The radius of the circle is the distance between the center (7, 4) and the point (0, 4) on the circle. Since the two points have the same y coordinates, then the distance between them is simply obtained by subtracting the lower value of the x coordinate of the two points from the higher value of the x coordinates of the two points. This is obtained as follows:

$r = 7 - 0$

$r = 7$

This means that when a circle touches the y-axis, the radius of the circle is a, which is the x coordinate of the center of the circle.

Hence, with center (7, 4) and radius 7, the equation of the circle is obtained as follows:

$(x - a)^2 + (y - b)^2 = r^2$

$(x - 7)^2 + (y - 4)^2 = 7^2$

$(x - 7)^2 + (y - 4)^2 = 49$

$x^2 - 14x + 49 + y^2 - 8y + 16 = 49$

$x^2 + y^2 - 14x - 8y + 16 + 49 - 49 = 0$

$x^2 + y^2 - 14x - 8y + 16 = 0$

13. A circle touches the positive x and y axes. If the line $5x - 3y - 18 = 0$ passes through the center of the circle, find the equation of the circle.

Solution

When a circle touches the x and y axes, then the coordinates of the center of the circle are the same, and they are equal to the radius of the circle. Hence, (a, b) = (r, r), where r is the radius of the circle. Since the line $5x - 3y - 18 = 0$ passes the center (r, r) of the circle, then it follows that the value of r can be obtained by putting $x = r$ and $y = r$ in the equation of the line. This is done as follows:

$5x - 3y - 18 = 0$

$5r - 3r - 18 = 0$

$2r = 18$

$r = 9$

Hence, the circle has center (9, 9) and radius 9. The equation of the circle is given by:

$(x - a)^2 + (y - b)^2 = r^2$

$(x-9)^2 + (y-9)^2 = 9^2$
$x^2 - 18x + 81 + y^2 - 18y + 81 = 81$
$x^2 + y^2 - 18x - 18y + 81 + 81 - 81 = 0$
$x^2 + y^2 - 18x - 18y + 81 = 0$

14. A circle has center $(-5, 4)$. If it touches the line $4x + 2y - 4 = 0$, find:
(a) the radius of the circle
(b) the equation of the circle
Solution
(a) Since the circle touches the line $2x + y - 2 = 0$, then the radius of the circle is the distance from the center of the circle to the line $2x + y - 2 = 0$. This is obtained as follows:
Recall that the perpendicular distance between a point (x_1, y_1) and a line $ax + by + c = 0$, is given by:

$$D = \frac{|ax_1 + by_1 + c|}{\sqrt{a^2 + b^2}}$$

Hence, the distance from the center $(-5, 4)$, of the circle to the line $2x + y - 2 = 0$ gives the radius as follows:

$$r = \frac{|ax_1 + by_1 + c|}{\sqrt{a^2 + b^2}}$$

$$= \frac{|2(-5) + 4 - 2|}{\sqrt{2^2 + 1^2}}$$

$$= \frac{|-10 + 4 - 2|}{\sqrt{5}}$$

$$= \frac{|-8|}{\sqrt{5}}$$

$$r = \frac{8}{\sqrt{5}} \qquad \text{(The bar sign means the negative sign should be ignored)}$$

Therefore, with center $(-5, 4)$ and radius $\frac{8}{\sqrt{5}}$, the equation of the circle is given by:

$$(x - a)^2 + (y - b)^2 = r^2$$

$$(x - (-5))^2 + (y - 4)^2 = \left(\frac{8}{\sqrt{5}}\right)^2$$

$$(x + 5)^2 + (y - 4)^2 = \frac{64}{5}$$

$$x^2 + 10x + 25 + y^2 - 8y + 16 = \frac{64}{5}$$

$$x^2 + y^2 + 10x - 8y + 25 + 16 - \frac{64}{5} = 0$$

Multiplying each term by 5 in order to clear out the fraction gives:
$$5x^2 + 5y^2 + 50x - 40y + 125 + 80 - 64 = 0$$
$$5x^2 + 5y^2 + 50x - 40y + 141 = 0$$

73

15. Show that the line $3x + 2y = 0$ touches the circle $x^2 + y^2 + 6x + 4y = 0$.

Solution

When a line touches a circle, the equation of the line and circle are solved simultaneously. The quadratic equation formed gives repeated (equal) roots. This shows that the circle touches the line once.

Let us now solve the two equations simultaneously as follows.

$$3x + 2y = 0 \ldots\ldots\ldots\ldots(1)$$
$$x^2 + y^2 + 6x + 4y = 0 \ldots\ldots\ldots(2)$$

From equation (1), $x = \dfrac{-2y}{3} \ldots\ldots\ldots(3)$

Substitute $\dfrac{-2y}{3}$ for x in equation (2)

$$x^2 + y^2 + 6x + 4y = 0 \ldots\ldots\ldots(2)$$
$$\left(\dfrac{-2y}{3}\right)^2 + y^2 + 6\left(\dfrac{-2y}{3}\right) + 4y = 0$$
$$\left(\dfrac{4y^2}{9}\right) + y^2 - 4y + 4y = 0$$

Multiplying each term by 9 gives:

$$4y^2 + 9y^2 - 36y + 36y = 0$$
$$13y^2 = 0$$
$$y^2 = 0$$
$$y = 0$$

We now find x by putting $y = 0$ in equation (3) as follows:

$$x = \dfrac{-2y}{3} \ldots\ldots\ldots(3)$$
$$x = \dfrac{-2(0)}{3}$$
$$x = 0$$

Hence, we have only a single point of contact which is (0, 0). This single point of contact shows that the line touches the circle.

16. Determine the equations of the tangents to the circle $x^2 + y^2 - 4x - 6y + 8 = 0$ which are parallel to the line $y = 2x$.

Solution

$$x^2 + y^2 - 4x - 6y + 8 = 0$$

General form: $x^2 + y^2 - 2ax - 2by + c = 0$

By comparison: $-2a = -4$

$$a = \dfrac{-4}{-2}$$
$$a = 2$$

Also, $-2b = -6$

$$b = \dfrac{-6}{-2}$$
$$b = 3$$

The radius of the circle is given by:

$$r^2 = a^2 + b^2 - c$$

$$r^2 = 2^2 + 3^2 - 8$$
$$= 4 + 9 - 8$$
$$r^2 = 5$$
$$r = \sqrt{5}$$

Hence, center (a, b) of the circle is (2, 3), while the radius is $\sqrt{5}$

The equation of the line is y = 2x. Comparing with y = mx + c show that the gradient, m, of the line is 2. Hence, the gradient of the tangent to the circle is also 2, since they are parallel line. Let the equation of the tangent to the circle be, y = mx + c. Putting m = 2 gives the tangent as:

$$y = 2x + c$$

Or, y − 2x − c = 0

The distance between the center (2, 3) and the tangent y − 2x − c = 0 will be equal to the radius of the circle. This is calculated as follows:

$$r = \frac{|ax_1 + by_1 + c|}{\sqrt{a^2 + b^2}} \qquad \text{(Distance between a point and a line)}$$

$$r = \frac{|y_1 - 2x_1 - c|}{\sqrt{a^2 + b^2}} \qquad \text{(Note that y can also come before } x \text{ in this formula)}$$

$$\sqrt{5} = \frac{|1(3) - 2(2) - c|}{\sqrt{1^2 + (-2)^2}} \qquad \text{(Recall that the point is (2, 3) while the radius is } \sqrt{5}\text{)}$$

$$\sqrt{5} = \frac{|3 - 4 - c|}{\sqrt{1 + 4}}$$

$$\sqrt{5} = \frac{|-1 - c|}{\sqrt{5}}$$

$$\sqrt{5}(\sqrt{5}) = -1 - c$$

$$5 = |-1 - c| \qquad \text{(Note that } \sqrt{5}(\sqrt{5}) = 5\text{)}$$

$$(-1 - c) = 5 \quad \text{or} \quad -(-1 - c) = 5$$

(Note that $|x| = x$ or $-x$ which means that $|-1 - c| = (-1 - c)$ or $-(-1 - c)$ as shown above)

$$c = -1 - 5 \qquad \text{or} \quad 1 + c = 5$$
$$c = -6 \qquad \qquad \text{or} \quad c = 4$$

Recall that the required line is y − 2x − c = 0

Putting c = 4 and −6 gives two lines as follows:

$$y - 2x - 4 = 0 \text{ and } y - 2x + 6 = 0$$

Therefore the equations of the two parallel tangents are y − 2x − 4 = 0 and y − 2x + 6 = 0

17. The equation of a circle is $x^2 + y^2 - 6x - 4y + 4 = 0$. If it touches the y-axis, find the coordinates of the point of contact.

Solution

On the y-axis, the value of x is 0. Hence, we put $x = 0$ in the equation of the circle and find the value of y as follows:

$$x^2 + y^2 - 6x - 4y + 4 = 0$$
$$(0)^2 + y^2 - 6(0) - 4y + 4 = 0$$
$$y^2 - 4y + 4 = 0$$

Solving this quadratic equation should give us a repeated (equal) root as follows:

$(y - 2)(y - 2) = 0$

Hence, $y = 2$ (Repeated roots)

Therefore the point of contact is (0, 2)

18. Show that the line $y = 2x$ is a tangent to the circle $x^2 + y^2 - 8x - y + 5 = 0$. Determine the point of contact.

<u>Solution</u>

$y = 2x$ ....(1)

$x^2 + y^2 - 8x - y + 5 = 0$ ....(2)

Put $y = 2x$ in equation (2) as follows.

$x^2 + y^2 - 8x - y + 5 = 0$ ....(2)

$x^2 + (2x)^2 - 8x - (2x) + 5 = 0$

$x^2 + 4x^2 - 8x - 2x + 5 = 0$

$5x^2 - 10x + 5 = 0$

Dividing each term by 5 gives:

$x^2 - 2x + 1 = 0$

Solving this quadratic equation simultaneously gives:

$(x - 1)(x - 1) = 0$

Therefore, $x = 1$ (This is a repeated root)

Since x is a repeated root, it shows that the line is a tangent to the circle.

Recall that: $y = 2x$

$y = 2(1)$

$y = 2$

Hence, the point of contact is (1, 2)

19(a) Show that the circle $x^2 + y^2 + 4x - 6y + 8 = 0$ and the circle $x^2 + y^2 - 8x - 4 = 0$ touch externally.

(b) Determine the equation of the tangent to the larger circle which is parallel to the tangent at their point of contact.

<u>Solution</u>

The two circles are as shown below.

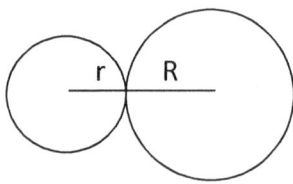

The diagram above makes it clear that when two circles touch externally, then the distance between their centers is equal to the sum of their radii.

Distance between their centers = R + r

Or, $D = R + r$

Let us now find the centers and radii as follows:

First equation: $x^2 + y^2 + 4x - 6y + 8 = 0$

General form: $x^2 + y^2 - 2ax - 2by + c = 0$

Comparing them, shows that:

$$-2a = 4$$
$$a = \frac{4}{-2}$$
$$a = -2$$

Also, $-2b = -6$

$$b = \frac{-6}{-2}$$
$$b = 3$$

Recall that: $r^2 = a^2 + b^2 - c$

$$r^2 = (-2)^2 + 3^2 - 8$$
$$= 4 + 9 - 8$$
$$r^2 = 5$$
$$r = \sqrt{5}$$

Therefore the center and radius of the first circle are $(-2, 3)$ and $\sqrt{5}$ respectively

Second equation: $x^2 + y^2 - 8x - 4 = 0$

General form: $x^2 + y^2 - 2ax - 2by + c = 0$

Comparing them, shows that:

$$-2a = -8$$
$$a = \frac{-8}{-2}$$
$$a = 4$$

Also, $-2b = 0$ (Since there is no term in y we equate $-2b$ to 0)

$$b = 0$$

$$r^2 = a^2 + b^2 - c$$
$$r^2 = (4)^2 + 0^2 - (-4)$$
$$= 16 + 0 + 4$$
$$r^2 = 20$$
$$r = \sqrt{20}$$

Therefore the center and radius of the second circle are $(4, 0)$ and $\sqrt{20}$ respectively.

Therefore, $R + r = \sqrt{20} + \sqrt{5}$

$$= \sqrt{4}(\sqrt{5}) + \sqrt{5}$$
$$= 2\sqrt{5} + \sqrt{5}$$
$$R + r = 3\sqrt{5}$$

Let us now find the distance between the centers (i.e. $(-2, 3)$ and $(4, 0)$) of the circle by the use of formula as follows:

$$D = \sqrt{(x_2 - x_1)^2 + (y_2 - y_1)^2}$$
$$= \sqrt{(4 - (-2))^2 + (0 - 3)^2}$$

$$= \sqrt{(4+2)^2 + (-3)^2}$$
$$= \sqrt{36 + 9}$$
$$= \sqrt{45}$$
$$= \sqrt{9}(\sqrt{5})$$
$$= 3\sqrt{5}$$

This shows that: D = R + r = $3\sqrt{5}$

Therefore, the two circles touch externally

(My book, "Simplified Mathematics: Numbers and Numeration" talks more about surds)

(b) At the point of contact, the two equations are equal. This gives:
$$x^2 + y^2 + 4x - 6y + 8 = x^2 + y^2 - 8x - 4$$
$$x^2 - x^2 + y^2 - y^2 + 4x + 8x - 6y + 8 + 4 = 0$$
$$12x - 6y + 12 = 0$$

Dividing each term by 6 gives:
$$2x - y + 2 = 0$$
$$y = 2x + 2 \dots\dots\dots(1)$$

Substitute $2x + 2$ for y in any of the two equations of the circles. This gives:
$$x^2 + y^2 - 8x - 4 = 0$$
$$x^2 + (2x + 2)^2 - 8x - 4 = 0$$
$$x^2 + 4x^2 + 8x + 4 - 8x - 4 = 0$$
$$5x^2 = 0$$
$$x = 0$$

Put $x = 0$ in equation (1) to find the value of y. This gives:
$$y = 2x + 2 \dots\dots\dots(1)$$
$$= 2(0) + 2$$
$$y = 2$$

Hence, the point of contact of the two circles is (0, 2)

The point of contact of the tangent to the larger circle is the opposite end of the diameter of the larger circle. This end and the common point of contact of the two circles give the midpoint (center) of the larger circle. Hence, we use the midpoint formula to find this end. Let the end be (m, n). The larger circle has its center at (4, 0), (since it has a larger radius of $\sqrt{20}$). With its ends of diameter as (0, 2) (i.e. the common point) and (m, n), we find the values of m and n by using the midpoint formula as follows:
$$x = \left(\frac{x_1 + x_2}{2}\right)$$

$$4 = \left(\frac{0 + m}{2}\right) \qquad \text{(Note that the center is (4, 0), with } x \text{ coordinate as 4)}$$
$$m = 8$$

Similarly:
$$y = \left(\frac{y_1 + y_2}{2}\right)$$

$$0 = \left(\frac{2+n}{2}\right) \qquad \text{(Note that the center is (4, 0), with y coordinate as 0)}$$
$$2 + n = 0$$
$$n = -2$$

Therefore, the other end of the diameter of the larger circle where the parallel tangent passes is (8, –2). With this point of contact, the equation of a tangent to a circle at the pint (x_1, y_1) is given by:
$$xx_1 + yy_1 - a(x + x_1) - b(y + y_1) + c = 0$$
Recall that the equation of the larger circle is $x^2 + y^2 - 8x - 4 = 0$. Hence we substitute the appropriate values into the equation above as follows:
$$x(8) + y(-2) - 4(x + 8) - 0(y - 2) - 4 = 0 \qquad \text{(Note that center (a, b) = (4, 0) and c = –4)}$$
$$8x - 2y - 4x - 32 + 0 - 4 = 0$$
$$4x - 2y - 36 = 0$$
Dividing each term by 2 gives:
$$2x - y - 18 = 0$$

20. If the line $3x - 4y - 7 = 0$ cuts the circle $x^2 + y^2 - 16x - 8y + 6 = 0$ at the points A and B, find the coordinates of the points A and B.

Solution

In order to find the points where the line cuts the circle, we have to solve the two equations simultaneously as follows:
$$3x - 4y - 7 = 0 \dots\dots\dots\dots(1)$$
$$x^2 + y^2 - 16x - 8y + 6 = 0 \dots\dots\dots\dots(2)$$
From equation (1), $x = \dfrac{4y+7}{3} \dots\dots\dots\dots(3)$

Substitute $\dfrac{4y+7}{3}$ for x in equation (2) as follows.
$$x^2 + y^2 - 16x - 8y + 6 = 0 \dots\dots\dots\dots(2)$$
$$\left(\frac{4y+7}{3}\right)^2 + y^2 - 16\left(\frac{4y+7}{3}\right) - 8y + 6 = 0$$

$$\frac{16y^2+56y+49}{9} + y^2 - \left(\frac{64y+112}{3}\right) - 8y + 6 = 0$$
Multiply throughout by 9 (i.e. the LCM of 9 and 3) in order to clear out the fractions. This gives:
$$16y^2 + 56y + 49 + 9y^2 - 3(64y + 112) - 72y + 54 = 0$$
$$16y^2 + 56y + 49 + 9y^2 - 192y - 336 - 72y + 54 = 0$$
$$25y^2 - 208y - 233 = 0$$
Let us now use the quadratic equation formula to solve this quadratic equation as follows:
$$y = \frac{-b \pm \sqrt{b^2 - 4ac}}{2a}$$
From the equation $25y^2 - 208y - 233 = 0$, a = 25, b = –208, c = –233.
$$y = \frac{-(-208) \pm \sqrt{(-208)^2 - 4(25)(-233)}}{2(25)}$$

79

$$= \frac{208 \pm \sqrt{43264 + 23300}}{100}$$

$$= \frac{208 \pm \sqrt{66564}}{100}$$

$$= \frac{208 \pm 258}{100}$$

$$y = \frac{208 + 258}{100} \quad \text{or} \quad y = \frac{208 - 258}{100}$$

$$y = \frac{466}{100} \quad \text{or} \quad y = \frac{-50}{100}$$

$$y = 4.66 \quad \text{or} \quad y = -0.5$$

When y = 4.66, we find x from equation (3) as follows:

$$x = \frac{4y + 7}{3} \quad \text{............(3)}$$

$$= \frac{4(4.66) + 7}{3}$$

$$= \frac{18.64 + 7}{3}$$

$$x = 8.55$$

Similarly, when y = −0.5, we find x from equation (3) as follows:

$$x = \frac{4y + 7}{3} \quad \text{............(3)}$$

$$= \frac{4(-0.5) + 7}{3}$$

$$= \frac{-2 + 7}{3}$$

$$x = 1.67$$

Therefore, the points A and B are (1.67, −0.5) and (8.55, 4.66)

21. The equation of a circle is $x^2 + y^2 - 10x + 8y = 0$. Find the length of its chord along the x axis.

Solution

On the x-axis, the value of y is zero. Hence, we substitute 0 for y in the equation of the circle in order to find the two points where the x-axis cuts the circle as a chord.

$$x^2 + y^2 - 10x + 8y = 0$$
$$x^2 + (0)^2 - 10x + 8(0) = 0$$
$$x^2 - 10x = 0$$

Solving this quadratic equation gives:

$$x(x - 10) = 0$$

Therefore, $x = 0$ or $x = 10$

This solution shows that on the x-axis when y = 0, $x = 0$ and 10. Hence, the two points where the chord (x-axis) cuts the circle are (0, 0) and (10, 0). The length of the chord is the distance between these two points which is obtained as shown below.

L = 10 − 0   (We simply subtract the lower x coordinate from the higher one since the y coordinates are equal)

L = 10

Therefore, the length of the chord is 10 units.

22. The coordinates of the vertices of a triangle ABC are A(3, 2), B(−1, −2) and C(5, −4). Find the equation of the circumcircle of triangle ABC.

Solution

The point of intersection of any two perpendicular bisectors of two sides of the triangle gives the center of the circle required.

The mid-point of line AB (A(3, 2), B(−1, −2)) is obtained as follows:

$$x = \left(\frac{x_1 + x_2}{2}\right)$$

$$= \left(\frac{3 - 1}{2}\right)$$

$$x = 1$$

And $y = \left(\frac{y_1 + y_2}{2}\right)$

$$= \left(\frac{2 - 2}{2}\right)$$

$$= 0$$

Hence, the midpoint of AB = (1, 0)

The gradient of AB, (3, 2) and (−1, −2) is obtained as follows.

$$m_1 = \frac{y_2 - y_1}{x_2 - x_1}$$

$$= \frac{-2 - 2}{-1 - 3}$$

$$= \frac{-4}{-4}$$

$$m_1 = 1$$

Hence, the gradient of the perpendicular bisector of AB is $m_2 = -1$ (since $m_1 m_2 = -1$).

We now use the gradient $m_2 = -1$ and the midpoint of AB, i.e. (1, 0) to find the equation of the perpendicular bisector of AB as follows:

$$m_2 = \frac{y - y_1}{x - x_1}$$

$$-1 = \frac{y - 0}{x - 1}$$

$$y = -1(x - 1)$$

$$y = -x + 1$$

$$y + x = 1 \dots\dots\dots(1)$$

We now carry out this whole process for line BC as follows.

The mid-point of line BC (B(−1, −2) and C(5, −4)) is obtained as follows:

$$x = \left(\frac{x_1 + x_2}{2}\right)$$

$$= \left(\frac{-1 + 5}{2}\right)$$

$$x = 2$$

And $y = \left(\frac{y_1 + y_2}{2}\right)$

$$= \left(\frac{-2 - 4}{2}\right)$$

$$= -3$$

Hence, the midpoint of BC = (2, –3)

The gradient of BC, B(–1, –2) and C(5, –4) is obtained as follows.

$$m_1 = \frac{y_2 - y_1}{x_2 - x_1}$$

$$= \frac{-4 - (-2)}{5 - (-1)}$$

$$= \frac{-2}{6}$$

$$m_1 = -\frac{1}{3}$$

Hence, the gradient of the perpendicular bisector of BC is $m_2 = 3$ (Either you solve, $m_1 m_2 = -1$ and find m_2, or you simply take the negative inverse of m_1 to get m_2).

We now use the gradient $m_2 = 3$ and the midpoint of BC, i.e. (2, –3) to find the equation of the perpendicular bisector of BC as follows:

$$m_2 = \frac{y - y_1}{x - x_1}$$

$$3 = \frac{y - (-3)}{x - 2}$$

$$3 = \frac{y + 3}{x - 2}$$

$$y + 3 = 3(x - 2)$$

$$y + 3 = 3x - 6$$

$$y - 3x = -9 \ \dots\dots\dots\dots(2)$$

Let us now solve the equation (1) and (2) simultaneously to find the center of the circle.

$$y + x = 1 \ \dots\dots\dots\dots(1)$$

$$y - 3x = -9 \ \dots\dots\dots\dots(2)$$

From equation (1), $x = 1 - y \ \dots\dots\dots\dots(3)$

Substitute $1 - y$ for x in equation (2) as follows:

$$y - 3x = -9 \ \dots\dots\dots\dots(2)$$

$$y - 3(1 - y) = -9$$

$$y - 3 + 3y = -9$$

$$4y = 3 - 9$$

$$y = \frac{-6}{4}$$

$$y = -\frac{3}{2}$$

We now find x from equation 3 as follows:

$$x = 1 - y \,............(3)$$

$$= 1 - (-\frac{3}{2})$$

$$= 1 + \frac{3}{2}$$

$$x = \frac{5}{2}$$

Therefore, the center of the circle is $\left(\frac{5}{2}, -\frac{3}{2}\right)$

The distance from the center of the circle to any of the points A, B, C, on the circumference of the circle gives the radius of the circle. Hence, using the point A(3, 2) and the center $\left(\frac{5}{2}, -\frac{3}{2}\right)$, to find the radius gives:

$$r^2 = (x_2 - x_1)^2 + (y_2 - y_1)^2$$

$$= (\frac{5}{2} - 3)^2 + (-\frac{3}{2} - 2)^2$$

$$= (\frac{5-6}{2})^2 + (\frac{-3-4}{2})^2$$

$$= (\frac{-1}{2})^2 + (\frac{-7}{2})^2$$

$$= \frac{1}{4} + \frac{49}{4}$$

$$= \frac{50}{4}$$

$$r^2 = \frac{25}{2}$$

Hence, with center $\left(\frac{5}{2}, -\frac{3}{2}\right)$ and $r^2 = \frac{25}{2}$ we obtain the equation of the circumcircle as follows:

$$(x - a)^2 + (y - b)^2 = r^2$$

$$(x - \frac{5}{2})^2 + (y - (-\frac{3}{2}))^2 = \frac{25}{2}$$

$$x^2 - 5x + \frac{25}{4} + y^2 + 3y + \frac{9}{4} - \frac{25}{2} = 0$$

Multiply each term by 4 to clear out the fractions. This gives:

$$4x^2 - 20x + 25 + 4y^2 + 12y + 9 - 50 = 0$$

$$4x^2 + 4y^2 - 20x + 12y + 25 + 9 - 50 = 0$$

$$4x^2 + 4y^2 - 20x + 12y - 16 = 0$$

Dividing each term by 4 gives:

$$x^2 + y^2 - 5x + 3y - 4 = 0$$

Therefore, the equation of the circumcircle of the triangle is $x^2 + y^2 - 5x + 3y - 4 = 0$.

Exercise 4

1. Find the equation of a circle of radius 9 units and center at the origin.
2. Determine the equation of the circle of center $(-1, -1)$ and radius 5 units
3. The center of a circle is at the point $(2, 5)$ and the circle passes through $(4, 1)$. Find:
(a) the radius of the circle
(b) the equation of the circle.
4. The coordinates of the ends of the diameter of a circle are given by $(3, -2)$ and $(-6, 5)$. Find:
(a) the radius of the circle
(b) the equation of the circle
5. The equation of a circle is given by: $x^2 + y^2 + 4x - 6y + 8 = 0$. Find the:
(a) centre
(b) radius of the circle.
6. Determine the center and radius of a circle whose equation is: $3x^2 + 3y^2 - 9x + 21y - 16 = 0$
7. Find the equation of the circle which lies on the points $(2, 1)$, $(1, -6)$ and $(5, 2)$.
8. A circle passes through the points $(1, 2)$, $(3, -4)$ and $(5, -6)$. Find:
(a) the radius of the circle
(b) the equation of the circle
9. The equation of a circle is $x^2 + y^2 + 5x - 8y - 2 = 0$. Find the equation of a straight line which passes through the center of the circle and is parallel to the line $x - 2y = 5$
10. Which of the following equations describe a circle? If not give reasons for your answers.
(a) $x^2 + y^2 = 10$
(b) $x^2 + 2y^2 + 11x - 15y = 0$
(c) $2x^2 + y^2 - 7x + 11y = 0$
(d) $5x^2 + 5y^2 + 9x + 14y = 0$
(e) $3x^2 + 3y^2 - 6x = 5$
(f) $3x^2 + 3y^2 - 5y - 8 = 0$
(g) $4x^2 + 4y^3 - 2xy + 11y - 15 = 0$
(h) $x^2 - y^2 - x + y = 1$
(i) $0.5x^2 + 0.5y^2 - 5.6x + 7.2y + 12 = 0$
11. Show that the point $(1, 2)$ lies on the circle: $x^2 + y^2 - 3x + 5y - 12 = 0$
12. The line $x + y - 3 = 0$ cuts the circle $x^2 + y^2 = 9$ at P and Q. Find the points P and Q.
13. $P(1, 1)$ and $Q(2, -2)$ are two points on an x-y plane. Point $A(x, y)$ moves in the plane such that $|PA| : |QA| = 4 : 3$. Determine the equation of the locus A
14. A point M moves in an x-y plane such that its distance from the origin O is thrice its distance from the point $(2, 1)$.
(a) Find the equation of the locus of M
(b) Find its center and radius
15. Find the equation of a circle whose center is $(2, -5)$ and circumference 10π.

16. The equation of a circle is given by:

$$x^2 + y^2 - 4x + 2y + k = 0$$

If the radius of the circle is 7 units, find the value of the constant k.

17. The equation of a circle is given by $3x^2 + 3y^2 - 12x - 6y + 10 = 0$. Find the:

(a) area of the circle

(b) length of the diameter of the circle

18. The equation of a circle is $x^2 + y^2 + 8x - 14y - 9 = 0$. Find the equation of a straight line which passes through the center of the circle and is parallel to the line $4x + 2y = 5$

19. The equation of a circle is $2x^2 + 2y^2 - 10x + 3y - 17 = 0$. Determine the equation of the tangent to the circle at the point (−1, 1) on the circle.

20. Find the equation of the tangent to the circle $x^2 + y^2 - 4x + 12y - 42 = 0$ at the point (1, 3).

21. The equation of a circle is $x^2 + y^2 + 5x + 10y - 5 = 0$. Determine the equation of the normal to the circle at the point (−2, 1) on the circle.

22. Find the equation of the normal to the circle $x^2 + y^2 - 4x - 2y - 27 = 0$ at the point (−2, −3).

23. Find the length of the tangent to circle $3x^2 + 3y^2 - 7x + 22y + 9 = 0$ from the point (12, −9)

24. Find the length of the tangent to the circle $x^2 + y^2 - 4x + 6y + 9 = 0$ from the point (5, 7).

25. The line $x + y + 3 = 0$ passes through the center of a circle. If the circle touched the x-axis at the point (−2, 0), find the equation of the circle.

26. The line $2x - y - 1 = 0$ passes through the center of a circle. If the circle touched the x-axis at the point (3, 0), find the equation of the circle.

27. A circle touches the y-axis at the point (0, 7). If the line $3x - 5y + 5 = 0$ passes through the center of the circle, find the equation of the circle.

28. A circle touches the positive x and y axes. If the line $2x + y - 15 = 0$ passes through the center of the circle, find the equation of the circle.

29. A circle has center (3, 5). If it touches the line $6x - 8y - 3 = 0$, find:

(a) the radius of the circle

(b) the equation of the circle

30. Show that the line $7y - x = 5$ touches the circle $x^2 + y^2 - 5x + 5y = 0$. Find the point of contact.

31. Determine the equations of the tangents to the circle $x^2 + y^2 - 6x - 10y + 24 = 0$ which are parallel to the line $y = x$.

32. The equation of a circle is $x^2 + y^2 - 2x - 6y + 4 = 0$. If it touches the y-axis, find the coordinates of the point of contact.

33. Show that the line $7x - 3y = 1$ is a tangent to the circle $x^2 + y^2 + 5x - 7y + 4 = 0$. Determine the point of contact.

34 Show that the circle $x^2 + y^2 + 2x - 2y - 7 = 0$ and the circle $x^2 + y^2 - 6x + 4y + 9 = 0$ touch externally. What is the sum of their radii?

35. If the line $y - x - 1 = 0$ cuts the circle $x^2 + y^2 + 18x + 20y + 81 = 0$ at the points P and Q, find

the coordinates of the points P and Q.

36. The equation of a circle is $x^2 + y^2 - 2x - y - 3 = 0$. Find the length of its chord along the x axis.

37. The coordinates of the vertices of a triangle LMN are L(2, 3), M(1, 2) and N(3, 1). Find the equation of the circumcircle of triangle LMN.

38. If the line $x = 1$ touches the circle $x^2 + y^2 + 18x + 20y + 81 = 0$, find the coordinates of the point of contact.

CHAPTER 5
THE PARABOLA

What is a Parabola

A simple way to picture how a parabola looks like is to recall the shape of a quadratic curve.
A parabola is a locus of points equidistant from a given point called the focus and from a given line called the directrix. The vertex is the sharp turning point of the parabola. The line through the focus and perpendicular to the directrix is the axis of the parabola. The point of intersection of this axis and the parabola is the vertex of the parabola. Also, the line segment which passes the focus and is perpendicular to the axis of symmetry is called the latus rectum.
The diagrams below show the four types of shapes of a parabola .

(a)

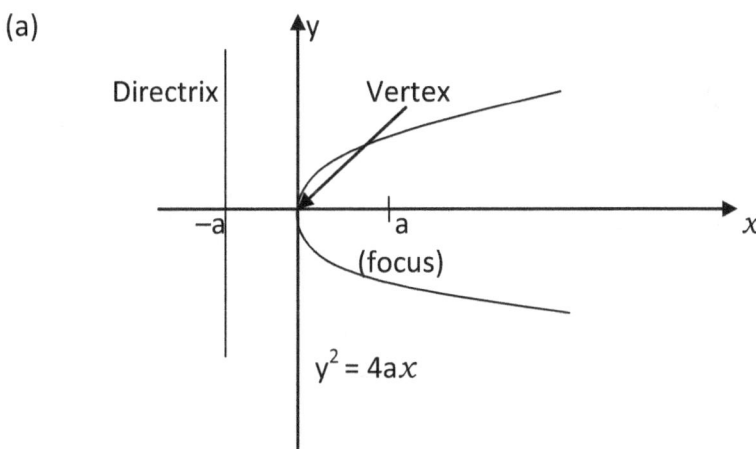

$y^2 = 4ax$

(b)

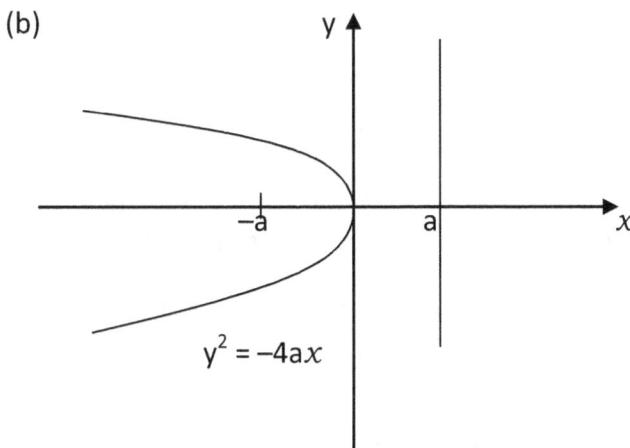

$y^2 = -4ax$

(c)

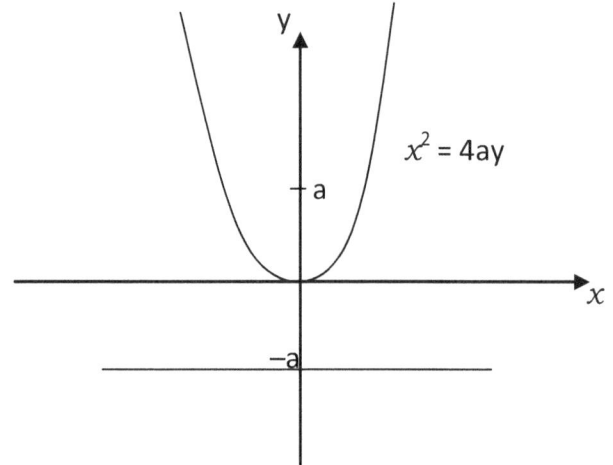

(d)

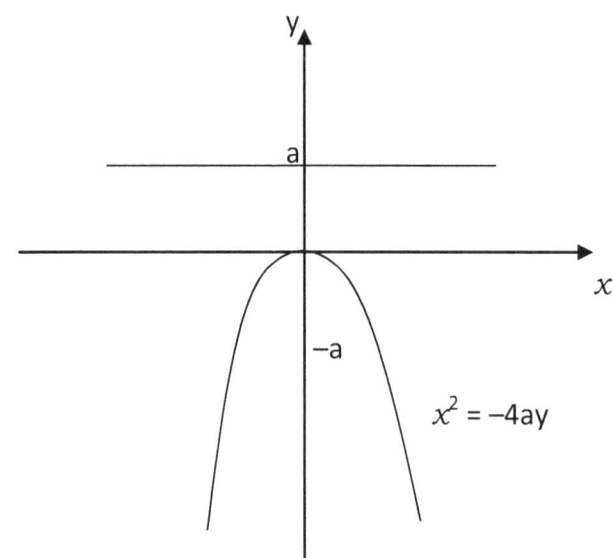

When the vertex of a parabola is at the origin (0, 0) and the axis of symmetry is along the x or y axis, then the equation of the parabola can be any of the four simplest forms stated below.

1. $y^2 = 4ax$. The focus will be at (a, 0), equation of directrix is: $x = -a$, the length of the latus rectum = 4a, and the parabola faces the positive x-axis (i.e. faces the right).

2. $y^2 = -4ax$. The focus will be at (−a, 0), equation of directrix is: $x = a$, the length of the latus rectum = 4a, and the parabola faces the negative x-axis (i.e. faces the left).

3. $x^2 = 4ay$. The focus will be at (0, a), equation of directrix is: $y = -a$, the length of the latus rectum = 4a, and the parabola faces the positive y-axis (i.e. faces upwards).

4. $x^2 = -4ay$. The focus will be at $(0, -a)$, equation of directrix is: $y = a$, the length of the latus rectum = 4a, and the parabola faces the negative y-axis (i.e. faces downwards).

Equation of a Parabola when the Vertex is not at the Origin

If the vertex of a parabola is translated from the origin to the point (x_1, y_1), then the equation of the corresponding parabola can be any of the four possible standard or canonical forms stated below.

1. $(y - y_1)^2 = 4a(x - x_1)$. Here, the focus is given by $(a + x_1, y_1)$, the equation of the directrix is: $x = -a + x_1$, the vertex is (x_1, y_1).

2. $(y - y_1)^2 = -4a(x - x_1)$. Here, the focus is given by $(-a + x_1, y_1)$, the equation of the directrix is: $x = a + x_1$, the vertex is (x_1, y_1).

3. $(x - x_1)^2 = 4a(y - y_1)$. Here, the focus is given by $(x_1, a + y_1)$, the equation of the directrix is: $y = -a + y_1$, the vertex is (x_1, y_1).

4. $(x - x_1)^2 = -4a(y - y_1)$. Here, the focus is given by $(x_1, -a + y_1)$, the equation of the directrix is: $y = a + y_1$, the vertex is (x_1, y_1).

Note that in all the four cases above, the length of the latus rectum is 4a.

Equation of the Tangent and Normal to the Parabola $y^2 = 4ax$ at point (x_1, y_1)

1. The equation of a tangent to a parabola $y^2 = 4ax$ at the point (x_1, y_1) is given by:
$$yy_1 = 2a(x + x_1)$$
The equation of a normal to a parabola $y^2 = 4ax$ at the point (x_1, y_1) is given by:
$$2ay + xy_1 = 2ay_1 + x_1y_1$$

2. The equation of a tangent to a parabola $y^2 = -4ax$ at the point (x_1, y_1) is given by:
$$yy_1 = -2a(x + x_1)$$

3. The equation of a tangent to a parabola $x^2 = 4ay$ at the point (x_1, y_1) is given by:
$$xx_1 = 2a(y + y_1)$$

4. The equation of a tangent to a parabola $x^2 = -4ay$ at the point (x_1, y_1) is given by:
$$xx_1 = -2a(y + y_1)$$

Equation of Tangent to a Parabola in the Slope Form

1. If the line $y = mx + c$ is a tangent to the parabola $y^2 = 4ax$, then the point of contact is given by: $(\frac{a}{m^2}, \frac{2a}{m})$, and $c = \frac{a}{m}$ which is the condition for tangency.

2. If the line $y = mx + c$ is a tangent to the parabola $y^2 = -4ax$, then the point of contact is given by: $(-\frac{a}{m^2}, -\frac{2a}{m})$, and $c = -\frac{a}{m}$ which is the condition for tangency

3. If the line $y = mx + c$ is a tangent to the parabola $x^2 = 4ay$, then the point of contact is given by: $(2am, am^2)$, and $c = -am^2$ which is the condition for tangency.

4. If the line $y = mx + c$ is a tangent to the parabola $x^2 = -4ay$, then the point of contact is given by: $(-2am, -am^2)$, and $c = am^2$ which is the condition for tangency.

Examples

1. Find the vertex, focus and directrix of the parabola, $y^2 = 20x$.
Solution
$\quad y^2 = 20x$
This equation is in the form: $y^2 = 4ax$. Equations in this form have vertex at the origin $(0, 0)$.
Equation: $\quad\quad y^2 = 20x$
Stabdard form: $y^2 = 4ax$
Comparing the two equations shows that:
$\quad\quad 4a = 20$
$\quad\quad a = \dfrac{20}{4}$
$\quad\quad a = 5$
Hence, the focus of this type of a parabola is given by: $(a, 0)$, which is $(5, 0)$
Also, its directrix is given by: $x = -a$
Therefore the equation of the directrix is $x = -5$

2. The focus of a parabola whose vertex is the origin is at the point $(-3, 0)$. Find:
(a) the equation
(b) the directrix
(c) the length of the latus rectum of the parabola.
Solution
(a) Since the vertex is the origin, it means that the focus is the point $(-a, 0)$
Hence, $(-a, 0) = (-3, 0)$
This shows that: $-a = -3$
$\quad\quad\quad\quad\quad a = 3$
Since the focus is $(-a, 0)$, it means that the form of the equation of the parabola is:
$\quad\quad y^2 = -4ax$
This means that: $y^2 = -4(3)x \quad$ (Since a = 3)
$\quad\quad y^2 = -12x \quad$ (This is the equation of the parabola)

(b) The equation of the directrix is given by:
$\quad\quad x = a$
This gives: $x = 3$

(c) The length of the latus rectum (LR) is given by:
$\quad\quad$ LR = 4a
$\quad\quad\quad$ = 4(3)

LR = 12 units

3. The equation of a parabola is given by: $x^2 = 48y$. Find:
(a) the focus
(b) the directrix of the parabola
(c) the direction that its open end is facing

Solution
(a) Equation: $x^2 = 48y$
Standard form: $x^2 = 4ay$
Comparing the two equations shows that:

$$4a = 48$$
$$a = \frac{48}{4}$$
$$a = 12$$

Recall that if a parabola is $x^2 = 4ay$, then its focus is the point $(0, a)$
Hence, focus, F = (0, 12)

(b) Recall that if a parabola is $x^2 = 4ay$, then the directrix is $y = -a$
Hence, directrix is: $y = -12$

(c) A parabola in the form $x^2 = 4ay$ opens upwards

4. The equation of a parabola is given by: $x^2 = -16y$. Find:
(a) the focus
(b) the directrix of the parabola

Solution
(a) Equation: $x^2 = -16y$
Standard form: $x^2 = -4ay$
Comparing the two equations shows that:

$$-4a = -16$$
$$a = \frac{-16}{-4}$$
$$a = 4$$

Recall that if a parabola is $x^2 = -4ay$, then its focus is the point $(0, -a)$
Hence, focus, F = (0, -4)

(b) If a parabola is $x^2 = -4ay$, then the directrix is $y = a$
Hence, directrix is: $y = 4$

5. The equation of a parabola is $y^2 - 6y - 2x + 19 = 0$. Express the equation in the canonical/standard form. Hence, determine:
(a) the vertex
(b) the focus

(c) the directrix of the parabola

Solution
$$y^2 - 6y - 2x + 19 = 0$$

In order to express it in the canonical form, simply complete the square of the quadratic part of the equation. This is as follows:
$$y^2 - 6y - 2x + 19 = 0$$
$$y^2 - 6y = 2x - 19$$

Divide the coefficient of y by 2. Square the value obtained and add it to both sides of the equation. When we divide the coefficient of y by 2 it gives:
$$\frac{-6}{2} = -3$$

Squaring this value obtained gives: $(-3)^2$

Hence, we add $(-3)^2$ to both sides of the equation to give:
$$y^2 - 6y + (-3)^2 = 2x - 19 + (-3)^2$$

Now express the left hand side as a square. In order to do this, simply take y and −3 and enclose them in a bracket and square the bracket. Do this, while simplifying the right hand side as follows:
$$(y - 3)^2 = 2x - 19 + 9$$
$$(y - 3)^2 = 2x - 10$$
$$(y - 3)^2 = 2(x - 5) \quad \text{(Canonical form)}$$

(a) We now compare this equation with the standard form as follows:
$$(y - 3)^2 = 2(x - 5)$$
Standard form: $(y - y_1)^2 = 4a(x - x_1)$
By comparison, the vertex (x_1, y_1) is:
Vertex = (5, 3)

(b) Also, by comparison, 4a = 2
$$a = \frac{2}{4}$$
$$a = \frac{1}{2}$$

Recall that when the equation of a parabola is $(y - y_1)^2 = 4a(x - x_1)$, then the focus is given by:
$$\text{Focus} = (a + x_1, y_1)$$
Hence, focus = $(\frac{1}{2} + 5, 3)$
$$\text{Focus} = (\frac{11}{2}, 3)$$

(c) Recall also that the directrix of this form of equation of a parabola is given by: $x = -a + x_1$
Hence, $x = -\frac{1}{2} + 5$
$$x = \frac{9}{2}$$

6. The equation of a parabola is given by $x^2 + 4x - 6y + 22 = 0$. Express the equation in the canonical form. Hence, determine:

(a) the vertex
(b) the focus
(c) the directrix of the parabola

Solution
$$x^2 + 4x - 6y + 22 = 0$$

In order to express it in the canonical form, simply complete the square of the quadratic part of the equation. This is as follows:

$$x^2 + 4x - 6y + 22 = 0$$
$$x^2 + 4x = 6y - 22$$

Divide the coefficient of x by 2. Square the value obtained and add it to both sides of the equation. When we divide the coefficient of x by 2 it gives:

$$\frac{4}{2} = 2$$

Squaring this value obtained gives: $(2)^2$

Hence, we add $(2)^2$ to both sides of the equation to give:
$$x^2 + 4x + (2)^2 = 6y - 22 + (2)^2$$

Now express the left hand side as a square. In order to do this, simply take x and 2 and enclose them in a bracket and square the bracket. Do this, while simplifying the right hand side as follows:

$$(x + 2)^2 = 6y - 22 + 4$$
$$(x + 2)^2 = 6y - 18$$
$$(x + 2)^2 = 6(y - 3) \quad \text{(Canonical form)}$$

(a) We now compare this equation with the standard form as follows:
$$(x + 2)^2 = 6(y - 3)$$
Standard form: $(x - x_1)^2 = 4a(y - y_1)$
By comparison, the vertex (x_1, y_1) is:
Vertex = (−2, 3)   (Note that by comparison $−x_1 = +2$, hence, $x_1 = −2$. The same applies to y_1)

(b) Also, by comparison, $4a = 6$
$$a = \frac{6}{4}$$
$$a = \frac{3}{2}$$

Recall that when the equation of a parabola is $(x - x_1)^2 = 4a(y - y_1)$, then the focus is given by:
Focus = $(x_1, \ a + y_1)$
Hence, focus = $(-2, \frac{3}{2} + 3)$
Focus = $(-2, \frac{9}{2})$

(c) Recall also that the directrix of this form of equation of a parabola is given by: $y = −a + y_1$
Hence, $y = -\frac{3}{2} + 3$
$$y = \frac{3}{2}$$

93

7. Find the equation of the tangent to the parabola $y^2 = -12x$ at the point $(-\frac{3}{4}, -3)$

<u>Solution</u>

Method 1

The equation of the parabola is in the form:

$$y^2 = -4ax$$

Equation: $y^2 = -12x$

By comparison we obtain a as follows:

$$-4a = -12$$

$$a = \frac{-12}{-4}$$

$$a = 3$$

The equation of a tangent to a parabola in the form $y^2 = -4ax$ is given by:

$$yy_1 = -2a(x + x_1)$$

We now substitute 3 for a and $(-\frac{3}{4}, -3)$ for (x_1, y_1) as follows:

$$yy_1 = -2a(x + x_1)$$

$$y(-3) = -2(3)(x + (-\frac{3}{4}))$$

$$-3y = -6(x - \frac{3}{4})$$

$$-3y = -6x + \frac{9}{2}$$

Multiply each term by 2 in order to clear out the fraction. This gives:

$$-6y = -12x + 9$$

$$-2y = -4x + 3 \quad \text{(After dividing each term by 3)}$$

$$4x - 2y - 3 = 0$$

Method 2

For a parabola in the form $y^2 = -4ax$, the point of contact of the tangent, $y = mx + c$ is given in the slope form as: $(-\frac{a}{m^2}, -\frac{2a}{m})$, where $c = -\frac{a}{m}$

Hence, $(-\frac{a}{m^2}, -\frac{2a}{m}) = (-\frac{3}{4}, -3)$ [i.e. point (x_1, y_1)]

By comparing any of the corresponding terms in the brackets above, we can obtain m as follows:

$$-\frac{2a}{m} = -3$$

From method 1, a = 3. Hence, put a = 3 in the equation above. This gives:

$$-\frac{2(3)}{m} = -3$$

$$-\frac{2(3)}{m} = -3$$

$$m = \frac{-6}{-3}$$

$$m = 2$$

Also, $c = \frac{-a}{m}$

$$c = -\frac{3}{2} \qquad \text{(Since a = 3 and m = 2)}$$

Hence, the equation of the tangent is given by:

$$y = mx + c$$

$$y = 2x - \frac{3}{2}$$

Multiply each term by 2. This gives:

$$2y = 4x - 3$$
$$2y - 4x + 3 = 0$$

8. The equation of a parabola is given by $y^2 = 18x$. Find the equation of the:
(a) tangent and
(b) normal at the point (2, 6)

Solution

(a) The equation of the parabola is in the form:

$$y^2 = 4ax$$

Equation: $y^2 = 18x$

By comparison we obtain 'a' as follows:

$$4a = 18$$
$$a = \frac{18}{4}$$
$$a = \frac{9}{2}$$

The equation of a tangent to a parabola in the form $y^2 = 4ax$ is given by:

$$yy_1 = 2a(x + x_1)$$

We now substitute $\frac{9}{2}$ for a and (2, 6) for (x_1, y_1) as follows:

$$yy_1 = 2a(x + x_1)$$
$$y(6) = 2(\tfrac{9}{2})(x + 2)$$
$$6y = 9(x + 2)$$
$$6y = 9x + 18$$
$$2y = 3x + 6 \qquad \text{(After dividing each term by 3)}$$
$$2y - 3x - 6 = 0$$

(b) The equation of the normal is given by:

$$2ay + xy_1 = 2ay_1 + x_1y_1$$
$$2(\tfrac{9}{2})y + x(6) = 2(\tfrac{9}{2})(6) + 2(6)$$
$$9y + 6x = 54 + 12$$
$$9y + 6x = 66$$
$$3y + 2x - 22 = 0 \qquad \text{(After dividing by 3)}$$

9. Find the equation of the tangent to the parabola $x^2 = 12y$ at the point (6, 3).

$$x^2 = 12y$$

This parabola is of the form $x^2 = 4ay$

Hence, comparing terms shows that:

$$4a = 12$$

$$a = \frac{12}{4}$$

$$a = 3$$

When an equation of a parabola is of the form $x^2 = 4ay$, then the equation of the tangent is given by:

$$xx_1 = 2a(y + y_1)$$

Substitute (6, 3) for (x_1, y_1) and 3 for 'a'. This gives:

$$xx_1 = 2a(y + y_1)$$

$$x(6) = 2(3)(y + 3)$$

$$6x = 6y + 18$$

$$x = y + 3 \qquad \text{(After dividing each term by 6)}$$

Therefore, the equation of the tangent is $x - y - 3 = 0$

Equation of a Parabola given Focus and Directrix, and Vertex not at the Origin

The midpoint of the focus and the directrix gives the vertex of a parabola. Also, 'a' is half the distance between the focus and the directrix.

1. If the coordinates of the focus of a parabola is (x_f, y_f), and the directrix is $x = x_d$, then the parabola opens to the right if $x_f > x_d$. In this case, the coordinates of the vertex x_1 and y_1 are given by:

$$x_1 = \frac{x_f + x_d}{2}, \text{ and } y_1 = y_f$$

The value 'a' is given by:

$$a = \frac{|x_f - x_d|}{2}$$

Hence, the equation of the parabola is given by:

$$(y - y_1)^2 = 4a(x - x_1)$$

2. If all the conditions in (1) above hold, except that $x_f < x_d$, then the parabola opens to the left, and its equation is given by:

$$(y - y_1)^2 = -4a(x - x_1)$$

3. If the focus of a parabola is (x_f, y_f), and the directrix is $y = y_d$, then the parabola opens upwards if $y_f > d_d$. In this case, the coordinates of the vertex x_1 and y_1 are given by:

$$x_1 = x_f \text{ and } y_1 = \frac{y_f + y_d}{2},$$

The value 'a' is given by:

$$a = \frac{|y_f - y_d|}{2}$$

Hence, the equation of the parabola is given by:

$$(x - x_1)^2 = 4a(y - y_1)$$

4. If all the conditions in (3) above hold, except that $y_f < y_d$, then the parabola opens downwards, and its equation is given by:

$$(x - x_1)^2 = -4a(y - y_1)$$

Equation of a Parabola when Given the Focus and the Vertex

1. If the focus, (x_f, y_f) and the vertex (x_1, y_1) of a parabola are given, then the parabola is a horizontal one if $y_f = y_1$. If $x_f > x_1$, then the parabola opens to the right and its equation is given by:

$$(y - y_1)^2 = 4a(x - x_1)$$

where $a = x_f - x_1$ and the directrix is $x = 2x_1 - x_f$.
Similarly, if $x_f < x_1$, then the parabola opens to the left and its equation is given by:

$$(y - y_1)^2 = -4a(x - x_1)$$

where $a = x_1 - x_f$ and the directrix is $x = 2x_1 - x_f$.

2. If the focus, (x_f, y_f) and the vertex (x_1, y_1) of a parabola are given, then the parabola is a vertical one if $x_f = x_1$. If $y_f > y_1$, then the parabola opens upwards and its equation is given by:

$$(x - x_1)^2 = 4a(y - y_1)$$

where $a = y_f - y_1$ and the directrix is $y = 2y_1 - y_f$.
Similarly, if $y_f < y_1$, then the parabola opens downwards and its equation is given by:

$$(x - x_1)^2 = -4a(y - y_1)$$

where $a = y_1 - y_f$ and the directrix is $y = 2y_1 - y_f$.

Equation of a Parabola when given the Vertex and the Directrix

1. If (x_1, y_1) is the vertex and $x = x_d$ is the directrix of a parabola, then the parabola is a horizontal parabola. If $x_1 > x_d$, then the parabola opens to the right, and $a = x_1 - x_d$. Hence, the equation of the parabola is given by:

$$(y - y_1)^2 = 4a(x - x_1)$$

The focus is given by: $(2x_1 - x_d, y_1)$
Similarly, if $x_1 < x_d$, then the parabola opens to the left, and $a = x_d - x_1$. Hence, the equation of the parabola is given by:

$$(y - y_1)^2 = -4a(x - x_1)$$

The focus is given by: $(2x_1 - x_d, y_1)$

2. If (x_1, y_1) is the vertex and $y = y_d$ is the directrix of a parabola, then the parabola is a vertical parabola. If $y_1 > y_d$, then the parabola opens upwards, and $a = y_1 - y_d$. Hence, the equation of the parabola is given by:

$$(x - x_1)^2 = 4a(y - y_1)$$

The focus is given by: $(x_1, 2y_1 - y_d)$
Similarly, if $y_1 < y_d$, then the parabola opens downward, and $a = y_d - y_1$. Hence, the equation of the parabola is given by:

$$(x - x_1)^2 = -4a(y - y_1)$$

The focus is given by: $(x_1, 2y_1 - y_d)$

Examples

1. Find the equation of a parabola of focus (3, −2) and directrix, $x = -5$

<u>Solution</u>

The Focus, (x_f, y_f) is (3, −2) and the directrix, $x = x_d$ is $x = -5$. This shows that $x_f = 3$ and $y_f = -2$, and $x_d = -5$.

Comparing x_f and x_d shows that $x_f > x_d$. Since 3 > −5. (Parabola opens to the right).

Hence, the coordinates of the vertex, x_1 is given by:

$$x_1 = \frac{x_f + x_d}{2}$$
$$= \frac{3 - 5}{2}$$
$$x_1 = -1$$

And, $y_1 = y_f$

Hence, $y_1 = -2$

Therefore, the vertex is at the point (−1, −2)

Also, $a = \frac{|x_f - x_d|}{2}$

$$= \frac{|3 - (-5)|}{2} \qquad \text{(Note that a should be positive)}$$
$$= \frac{8}{2}$$
$$a = 4$$

Hence, the equation of the parabola is given by:

$$(y - y_1)^2 = 4a(x - x_1)$$
$$(y - (-2))^2 = 4(4)(x - (-1))$$
$$(y + 2)^2 = 16(x + 1)$$
$$y^2 + 4y + 4 = 16x + 16$$
$$y^2 + 4y - 16x + 4 - 16 = 0$$
$$y^2 + 4y - 16x - 12 = 0$$

2. The focus and directrix of a parabola are (−1, 4) and y = 6 respectively. Find the equation of the parabola.

<u>Solution</u>

The Focus, (x_f, y_f) is (−1, 4) and the directrix, $y = y_d$ is $y = 6$. This shows that $x_f = -1$ and $y_f = 4$, and $y_d = 6$.

Comparing y_f and y_d shows that $y_f < y_d$. (Parabola opens downward).

Hence, the coordinates of the vertex, y_1 is given by:

$$y_1 = \frac{y_f + y_d}{2}$$
$$= \frac{4 + 6}{2}$$
$$y_1 = 5$$

And, $x_1 = x_f$

Hence, $x_1 = -1$

Therefore, the vertex is at the point (−1, 5)

Also, $a = \dfrac{|y_f - y_d|}{2}$

$= \dfrac{|4 - 6|}{2}$ (Note that a should be positive)

$= \dfrac{|-2|}{2}$

$= \dfrac{2}{2}$ (Ignore the negative sign due to the bar lines)

$a = 1$

Hence, the equation of the parabola is given by:

$(x - x_1)^2 = -4a(y - y_1)$ (This is for a parabola that faces downwards)

$(x - (-1))^2 = -4(1)(y - 5)$

$(x + 1)^2 = 4(y - 5)$

$x^2 + 2x + 1 = 4y - 20$

$x^2 + 2x - 4y + 1 + 20 = 0$

$x^2 + 2x - 4y + 21 = 0$

3. The vertex of a parabola is given by (2, 3). If its focus is at (−1, 3), find the equation of the parabola and its directrix.

Solution

The focus, (x_f, y_f) is (−1, 3) and the vertex (x_1, y_1) is (2, 3). Since, $y_f = y_1$, it means that the parabola is a horizontal one. Also, $x_f < x_1$. Hence, the parabola opens to the left. Therefore:

$a = x_1 - x_f$

$= 2 - (-1)$

$a = 3$

Hence, the equation of the parabola is given by:

$(y - y_1)^2 = -4a(x - x_1)$ (For a parabola that opens to the left)

$(y - 3)^2 = -4(3)(x - 2)$

$y^2 - 6y + 9 = -12(x - 2)$

$y^2 - 6y + 9 = -12x + 24$

$y^2 - 6y + 12x + 9 - 24 = 0$

$y^2 - 6y + 12x - 15 = 0$

The equation of the directrix is given by:

$x = 2x_1 - x_f$

$= 2(2) - (-1)$

$= 4 + 1$

$x = 5$

4. The focus and vertex of a parabola are (−2, 1) and (−2, −3) respectively. Find the:

(a) equation of the parabola

(b) equation of the directrix of the parabola.

Solution

(a) The focus, (x_f, y_f) is (−2, 1) and the vertex (x_1, y_1) is (− 2, −3). Since, $x_f = x_1$, it means that the

99

parabola is a vertical one. Also, $y_f > y_1$. Hence, the parabola opens upwards. Therefore:

$$a = y_f - y_1$$
$$= 1 - (-3)$$
$$a = 4$$

Hence, the equation of the parabola is given by:

$$(x - x_1)^2 = 4a(y - y_1) \quad \text{(For a parabola that opens upwards)}$$
$$(x - (-2))^2 = 4(4)(y - (-3))$$
$$(x + 2)^2 = 16(y + 3)$$
$$x^2 + 4x + 4 = 16y + 48$$
$$x^2 + 4x - 16y + 4 - 48 = 0$$
$$x^2 + 4x - 16y - 44 = 0$$

(b) The equation of the directrix is given by:

$$y = 2y_1 - y_f$$
$$= 2(-3) - 1$$
$$= -6 - 1$$
$$y = -7$$

5. The equation of the directrix of a parabola is $y = -2$. If the vertex of the parabola is at the point $(-3, -4)$, find:
(a) the equation of the parabola
(b) the focus of the parabola

Solution

(a) (x_1, y_1) is $(-3, -4)$, and $y = y_d$ is $y = -2$. Hence, $x_1 = -3$, $y_1 = -4$ and $y_d = -2$. This is a vertical parabola since the directrix is in terms of y. Comparing y_1 and y_d shows that $y_1 < y_d$ ($-4 < -2$). Hence, the parabola opens downwards, and $a = y_d - y_1$.

Therefore, $a = -2 - (-4)$
$$= -2 + 4$$
$$a = 2$$

Hence, the equation of the parabola is given by:

$$(x - x_1)^2 = -4a((y - y_1)$$
$$(x - (-3))^2 = -4(2)(y - (-4))$$
$$(x + 3)^2 = -8(y + 4)$$
$$x^2 + 6x + 9 = -8y - 32$$
$$x^2 + 6x + 8y + 9 + 32 = 0$$
$$x^2 + 6x + 8y + 41 = 0$$

(b) The focus in this case is given by:

$$(x_1, 2y_1 - y_d)$$
$$= (-3, \ 2(-4) - (-2))$$
$$= (-3, \ -8 + 2)$$
$$= (-3, \ -6)$$

6. The vertex and directrix of a parabola are (4, 5) and $x = 1$ respectively. Determine:
(a) the equation of the parabola
(b) the focus of the parabola

Solution

(a) (x_1, y_1) is (4, 5), and $x = x_d$ is $x = 1$. Hence, $x_1 = 4$, $y_1 = 5$ and $x_d = 1$. This is a horizontal parabola since the directrix is in terms of x. Comparing x_1 and x_d shows that $x_1 > x_d$ (4 > 1). Hence, the parabola opens to the right, and a $= x_1 - x_d$.

Therefore, a $= 4 - 1$

$\quad\quad\quad$ a $= 3$

Hence, the equation of the parabola is given by:

$\quad (y - y_1)^2 = 4a((x - x_1)$
$\quad (y - 5)^2 = 4(3)(x - 4)$
$\quad (y - 5)^2 = 12(x - 4)$
$\quad y^2 - 10y + 25 = 12x - 48$
$\quad y^2 - 10y - 12x + 25 + 48 = 0$
$\quad y^2 - 10y - 12x + 73 = 0$

(b) The focus in this case is given by:

$\quad (2x_1 - x_d, \ y_1)$
$\quad = (2(4) - 1, \ 5)$
$\quad = (8 - 1, \ 5)$
$\quad = (7, \ 5)$

Exercise 5

1. Find the vertex, focus and directrix of the parabola, $y^2 = 24x$.
2. The focus of a parabola whose vertex is the origin is at the point (−5, 0). Find:
(a) the equation
(b) the directrix
(c) the length of the latus rectum of the parabola.
3. The equation of a parabola is given by: $x^2 = 32y$. Find:
(a) the focus
(b) the directrix of the parabola
(c) the direction that its open end is facing
4. The equation of a parabola is given by: $x^2 = -4y$. Find:
(a) the focus
(b) the directrix of the parabola
5. The equation of a parabola is $y^2 - 10y - 8x + 1 = 0$. Express the equation in the canonical/standard form. Hence, determine:
(a) the vertex
(b) the focus

(c) the directrix of the parabola

6. The equation of a parabola is given by $x^2 + 2x - 12y + 37 = 0$. Express the equation in the canonical form. Hence, determine:

(a) the vertex

(b) the focus

(c) the directrix of the parabola

7. Find the equation of the tangent to the parabola $y^2 = -20x$ at the point $(-\frac{5}{4}, 5)$

8. The equation of a parabola is given by $y^2 = 32x$. Find the equation of the:

(a) tangent and

(b) normal at the point (2, 8)

9. Find the equation of the tangent to the parabola $x^2 = -20y$ at the point (10, -5).

10. Find the equation of a parabola of focus (2, -5) and directrix, $x = -1$

11. The focus and directrix of a parabola are (2, 7) and y = 9 respectively. Find the equation of the parabola.

12. The vertex of a parabola is given by (1, -2). If its focus is at (4, -2), find the equation of the parabola and its directrix.

13. The focus and vertex of a parabola are (5, -2) and (5, 3) respectively. Find the:

(a) equation of the parabola

(b) equation of the directrix of the parabola.

14. The equation of the directrix of a parabola is y = -3. If the vertex of the parabola is at the point (2, 1), find:

(a) the equation of the parabola

(b) the focus of the parabola

15. The vertex and directrix of a parabola are (-5, -3) and $x = -2$ respectively. Determine:

(a) the equation of the parabola

(b) the focus of the parabola

CHAPTER 6
THE ELLIPSE

What is an Ellipse

An ellipse is the locus of a point which moves such that the sum of its distances from two fixed points is a constant value. The two fixed points are called the foci. The diagrams below show the two possible ways an ellipse can appear.

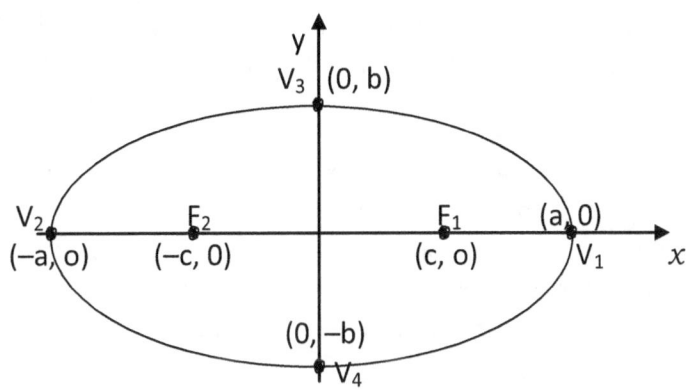

Ellipse with major axis horizontal

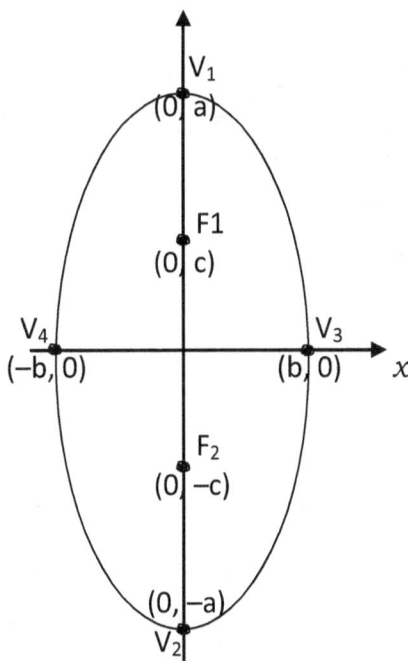

Ellipse with major axis vertical

If the major axis of an ellipse is horizontal, and its center is the origin (0, 0), then the standard or canonical form of the equation of the ellipse is given by:

$$\frac{x^2}{a^2} + \frac{y^2}{b^2} = 1 \qquad (a > b)$$

where a is half the length of the major axis (x-axis) and b is half the length of the minor axis (y-axis). The distance between the foci is 2c, where c is the distance between a focus and the center of the ellipse. c is given by: $c = \sqrt{a^2 - b^2}$ which means that $a^2 = b^2 + c^2$. In this case, the foci are at the point (c, 0) and (−c, 0), the vertices on the major axis are $V_1(a, 0)$ and $V_2(−a, 0)$, while the vertices on the minor axis are $V_3(0, b)$ and $V_4(0, −b)$.

If the major axis of an ellipse is vertical (y-axis), and its center is the origin (0, 0), then the standard or canonical form of the equation of the ellipse is given by:

$$\frac{x^2}{b^2} + \frac{y^2}{a^2} = 1 \qquad (a > b)$$

where a is half the length of the major axis (y-axis) and b is half the length of the minor axis (x-axis). In this case, the foci are at the point (0, c) and (0, −c), the vertices on the major axis are $V_1(0, a)$ and $V_2(0, −a)$, while the vertices on the minor axis are $V_3(b, 0)$ and $V_4(−b, 0)$.

If the center of an ellipse is transformed to the point (x_1, y_1) in such a way that the axes are parallel to the x and y axes, then the new standard/canonical form of the equation of the ellipse is given by:

$$\frac{(x - x_1)^2}{a^2} + \frac{(y - y_1)^2}{b^2} = 1 \qquad \text{(Major axis is horizontal)}$$

where the vertices on the major axis are $V_1(a + x_1, y_1)$ and $V_2(−a + x, y_1)$, while the vertices on the minor axis are $V_3(x_1, b+y_1)$ and $V_4(x_1, −b + y_1)$. The foci of the ellipse are at $F_1(c + x_1, y_1)$ and $f_2(−c + x_1, y_1)$

2. $$\frac{(x - x_1)^2}{b^2} + \frac{(y - y_1)^2}{a^2} = 1 \qquad \text{(Major axis is vertical)}$$

where the vertices on the major axis are $V_1(x_1, a + y_1)$ and $V_2(x_1, −a + y_1)$, while the vertices on the minor axis are $V_3(b+x_1, y_1)$ and $V_4(−b + x_1, y_1)$. The foci of the ellipse are at $F_1(x_1, c + y_1)$ and $F_2(x_1, −c + y_1)$.

Latus Rectum of an Ellipse

The latus rectum of an ellipse is a straight line in an ellipse which is perpendicular to the major axis and it passes through the focus of the ellipse.
The length of the latus rectum of an ellipse is given by:

$$L = \frac{2b^2}{a}$$

Eccentricity of an ellipse

The ratio of the distance of a point on an ellipse from the focus, to its distance from the directrix (a fixed line) is a constant value. This constant ratio is called the eccentricity of an ellipse. It is given by:

$$e = \sqrt{1 - \frac{b^2}{a^2}}$$

Or $\quad e = \frac{c}{a}$

For a parabola, e = 1, for an ellipse, e < 1, while for a hyperbola e > 1

Area and Perimeter of an Ellipse

The area of an ellipse can be calculated by using the formula:

$\quad$ A = πab

The approximate value of the perimeter of an ellipse is can be calculated by using any of the following formulas:

1. P = π(a + b) $\quad$ (This is used when a and b are almost equal)

2. $P = π\sqrt{2(a^2 + b^2)}$

3. $P = π(\frac{3}{2}(a + b) - \sqrt{ab}$

Equation of the Tangent to an Ellipse at the point (x_1, y_1)

The equation of the tangent to the ellipse $\frac{x^2}{a^2} + \frac{y^2}{b^2} = 1$ at the point (x_1, y_1) is given by:

$$\frac{xx_1}{a^2} + \frac{yy_1}{b^2} = 1$$

Or, $\quad b^2 xx_1 + a^2 yy_1 = a^2 b^2$

Equation of the Normal to an Ellipse at the Pont (x_1, y_1)

The equation of the normal to the ellipse $\frac{x^2}{a^2} + \frac{y^2}{b^2} = 1$ at the point (x_1, y_1) is given by:

$$a^2 xy_1 - b^2 x_1 y = x_1 y_1 (a^2 - b^2)$$

Equation of an Ellipse given the Foci and the Vertices, and the Center not at the Origin (0, 0)

In this case, the steps involved are:

1. The major axis is parallel to the x-axis if the y-coordinates of the vertices and foci are the same. Hence, use the equation:

$$\frac{(x - x_1)^2}{a^2} + \frac{(y - y_1)^2}{b^2} = 1$$

Note that (x_1, y_1) is the center of the ellipse.

The major axis is parallel to the y-axis if the x-coordinates of the vertices and foci are the same. Hence, use the equation:

$$\frac{(x - x_1)^2}{b^2} + \frac{(y - y_1)^2}{a^2} = 1$$

2. Find the center of the ellipse, (x_1, y_1) by finding the mid-point of the vertices. The distance from this center to any of the vertices gives the length a. Square it to get a^2.

3. Find c by calculating the distance between the center and any of the foci. Find b^2 by using:
$$a^2 = b^2 + c^2$$

4. Substitute the values of a^2, b^2, x_1 and y_1 into the appropriate equation as stated in step 1 above.
Note that if the foci and center or the vertices and the center of an ellipse are given, the equation of the ellipse can be obtained by using steps similar to the ones given above.

Examples

1 Find the four vertices and the foci of an ellipse whose equation is: $\dfrac{x^2}{4} + \dfrac{y^2}{25} = 1$

Solution
In this case, the center of the ellipse is the origin (0, 0).
$$\frac{x^2}{4} + \frac{y^2}{25} = 1$$

Standard frm: $\dfrac{x^2}{b^2} + \dfrac{y^2}{a^2} = 1$ (Note that a or a^2 is always the greater value)
Hence, it follows by comparison that:
$$a^2 = 25$$
$$a = \sqrt{25}$$
$$a = \pm 5$$
And $b^2 = 4$
$$b = \sqrt{4}$$
$$b = \pm 2$$
From $\dfrac{y^2}{a^2}$ it shows that the y coordinates of the vertices on the y-axis is a, while from $\dfrac{x^2}{b^2}$ it shows that the x coordinates of the vertices on the x-axis is b. Also, from $\dfrac{y^2}{a^2}$, we can deduce that the major axis is on the y-axis. Note that any axis that carries a in the equation is the major axis.
Hence, V_1 = (0, 5) and V_2 = (0, –5), because the y coordinates will carry the value of a since it is the major axis. Similarly, V_3 = (2, 0) and V_4 = (–2, 0), because the x coordinates will carry the value of b since it is the minor axis.
Recall that $c^2 = a^2 - b^2$
Hence, $c = \sqrt{a^2 - b^2}$
$$= \sqrt{25 - 4}$$
$$c = \pm\sqrt{21}$$
Hence, the foci are F_1 = (0, c) and F_2 = (0, –c), which gives:
$$F_1 = (0, \sqrt{21}) \text{ and } F_2 = (0, -\sqrt{21})$$
Note that the foci are always on the major axis, which is the y-axis in this case.

2. Find the vertices and the foci of the ellipse whose equation is: $\dfrac{x^2}{25} + \dfrac{y^2}{16} = 1$
Solution

This is also a case where the center of the ellipse is the origin (0, 0).

$$\frac{x^2}{25} + \frac{y^2}{16} = 1$$

Standard form: $\frac{x^2}{a^2} + \frac{y^2}{b^2} = 1$ (Note that a or a^2 is always the greater value)

Hence, it follows by comparison that:

$a^2 = 25$

$a = \sqrt{25}$

$a = \pm 5$

And $b^2 = 16$

$b = \sqrt{16}$

$b = \pm 4$

From $\frac{x^2}{a^2}$ it shows that the x coordinates of the vertices on the x-axis is a, while from $\frac{y^2}{b^2}$ it shows that the y coordinates of the vertices on the y-axis is b. Also, from $\frac{x^2}{a^2}$, we can deduce that the major axis is on the x-axis. Note that any axis that carries a in the equation is the major axis. Hence we use the value of a to obtain the vertices on the major axis as, V_1 = (5, 0) and V_2 = (−5, 0). Similarly, V_3 = (0, 4) and V_4 = (0, −4).

Recall that $c^2 = a^2 - b^2$

Hence, $c = \sqrt{a^2 - b^2}$

$= \sqrt{25 - 16}$

$c = \pm\sqrt{9}$

$c = \pm 3$

Hence, the foci are F_1 = (c, 0) and F_2 = (−c, 0). The values of c have to be on the x-axis since it is the major axis. Therefore, the foci are:

F_1 = (3, 0) and F_2 = (−3, 0)

3. Find the values of a, b and c in the equation of the ellipse: $16x^2 + 25y^2 = 1600$

Solution

$16x^2 + 25y^2 = 1600$

Divide each term by 1600 in order to make the right hand side to become 1. This gives:

$$\frac{16x^2}{1600} + \frac{25y^2}{1600} = \frac{1600}{1600}$$

$$\frac{x^2}{100} + \frac{y^2}{64} = 1$$

Standard form: $\frac{x^2}{a^2} + \frac{y^2}{b^2} = 1$ (Note that a^2 is always the greater value)

Hence, it follows by comparison that:

$a^2 = 100$

$a = \sqrt{100}$

$a = 10$

And $b^2 = 64$

$b = \sqrt{64}$

$b = 8$

Recall that $c^2 = a^2 - b^2$

Hence, $c = \sqrt{a^2 - b^2}$

$\quad = \sqrt{100 - 64}$

$c = \sqrt{36}$

$c = 6$

Therefore, a = 10, b = 8 and c = 6

4. Find the length of the major and minor axes of the ellipse $\dfrac{x^2}{16} + \dfrac{y^2}{49} = 1$

solution

$$\frac{x^2}{16} + \frac{y^2}{49} = 1$$

Standard form: $\dfrac{x^2}{b^2} + \dfrac{y^2}{a^2} = 1$ (The major axis is the y-axis)

Hence, it follows by comparison that:

$\quad a^2 = 49$

$\quad a = \sqrt{49}$

$\quad a = 7$

And $b^2 = 16$

$\quad b = \sqrt{16}$

$\quad b = 4$

Hence, the length of the major axis (longer axis) is given by 2a. Recall that that a is half the length of the major axis

Therefore, length of major axis = 2(7)

$\quad = 14$ units

and, length of minor axis = 2b

$\quad = 2(4)$

$\quad = 8$ units.

5. Find the equation of an ellipse with vertices (0, $\pm$13) and foci (0, $\pm$12).

 Solution

Since the x coordinates of the vertices and foci are the same (i.e. 0) it shows that the major axis is on the y-axis . Hence, $V_1 = (0, a)$ and $V_2 = (0, -a)$. Also, $F_1 = (0, c)$ and $F_2 = (0, -c)$. Comparing these points with the coordinates given in the question shows that:

$\quad V_1 = (0, 13), V_2 = (0, -13), F_1 = (0, 12)$ and $F_2 = (0, -12)$

Hence, a = 13 and c = 12.

Recall that: $a^2 = b^2 + c^2$

$\quad b^2 = a^2 - c^2$

$\quad\quad = 13^2 - 12^2$

$\quad\quad = 169 - 144$

$b^2 = 25$

Note that $a^2 = 13^2$

Hence, $a^2 = 169$

Therefore, the equation of the ellipse is:

$$\frac{x^2}{b^2} + \frac{y^2}{a^2} = 1 \qquad \text{(Note that a is on the y-axis, the major axis)}$$

Therefore the equation of the ellipse is $\dfrac{x^2}{25} + \dfrac{y^2}{169} = 1$

6. Find the equation of an ellipse with vertices (± 5, 0) and foci (± 4, 0).

Solution

Since the y coordinates of the vertices and foci are the same (i.e. 0) it shows that the major axis is on the *x*-axis . Hence, the *x* coordinates of the vertices shows that a = 5 while the *x* coordinates of the foci shows that c = 4.

Recall that: $a^2 = b^2 + c^2$

$$b^2 = a^2 - c^2$$
$$= 5^2 - 4^2$$
$$= 25 - 16$$
$$b^2 = 9$$

Note that $a^2 = 5^2$

Hence, $a^2 = 25$

Therefore, the equation of the ellipse is:

$$\frac{x^2}{a^2} + \frac{y^2}{b^2} = 1 \qquad \text{(Note that a is on the x-axis, the major axis)}$$

Therefore the equation of the ellipse is $\dfrac{x^2}{25} + \dfrac{y^2}{9} = 1$

7. The length of the semi-major and semi-minor axes of an ellipse are 10 and 6 units respectively. Find the:

(a) eccentricity

(b) length of the latus rectum

(c) the area of the ellipse

Solution

(a) Semi-major axis is a. Hence, a = 10

Semi-minor axis is b. Hence, b = 6

Eccentricity, $e = \sqrt{1 - \dfrac{b^2}{a^2}}$

$$= \sqrt{1 - \frac{6^2}{10^2}}$$

$$= \sqrt{1 - \frac{36}{100}}$$

$$= \sqrt{\frac{64}{100}}$$

$$= \frac{8}{10}$$

$$e = \frac{4}{5}$$

(b) The length of the latus rectum is given by:

$$L = \frac{2b^2}{a}$$

$$= \frac{2(6)^2}{10}$$

$$= \frac{2(36)}{10}$$

$$= \frac{72}{10}$$

$$L = 7.2 \text{ units}$$

(c) Area of an ellipse is given by:

$$A = \pi ab$$

$$= 3.142(10)(6)$$

$$= 188.52 \text{ square units}$$

8. An ellipse has foci (± 8, 0) and eccentricity, e = $\frac{4}{5}$. Find the equation of the ellipse.

Solution

The foci is (± 8, 0). This shows that the major axis is on the x-axis since y = 0 on the foci. It also shows that the center of the ellipse is the origin (0, 0).

Recall that foci, (± 8, 0) = ($\pm c$, 0). Therefore, c = 8.

Eccentricity, e = $\frac{c}{a}$

$$\frac{4}{5} = \frac{8}{a}$$

$$4a = 5(8)$$

$$a = \frac{40}{4}$$

$$a = 10$$

Recall also that:

$$a^2 = b^2 + c^2$$
$$b^2 = a^2 - c^2$$
$$= 10^2 - 8^2$$
$$= 100 - 64$$
$$b^2 = 36$$

Note that $a^2 = 10^2$
$$a^2 = 100$$

Since the x-axis is the major axis, the equation will take the form:
$$\frac{x^2}{a^2} + \frac{y^2}{b^2} = 1$$

This gives: $\dfrac{x^2}{100} + \dfrac{y^2}{36} = 1$

9. Find the equation of the ellipse having major axis along the x-axis and passing through the points (2, –2) and (–3, 1).

Solution

Since the major axis is on the x-axis, then the equation takes the form:
$$\frac{x^2}{a^2} + \frac{y^2}{b^2} = 1 \quad \text{('a' goes under the major axis, i.e } x\text{-axis)}$$

Substituting the first point (2, –2) into the equation above gives:
$$\frac{2^2}{a^2} + \frac{(-2)^2}{b^2} = 1 \quad \text{(Since } x = 2 \text{ and } y = -2)$$
$$\frac{4}{a^2} + \frac{4}{b^2} = 1 \ \text{.....................(1)}$$

Similarly, substituting the point (–3, 1) gives:
$$\frac{(-3)^2}{a^2} + \frac{(1)^2}{b^2} = 1$$
$$\frac{9}{a^2} + \frac{1}{b^2} = 1 \ \text{.....................(2)}$$

Bringing equations (1) and (2) together gives:
$$\frac{4}{a^2} + \frac{4}{b^2} = 1 \ \text{.....................(1)}$$
$$\frac{9}{a^2} + \frac{1}{b^2} = 1 \ \text{.....................(2)}$$

Multiply equation (2) by 4 so that b^2 can be eliminated.

This gives: $4(\dfrac{9}{a^2} + \dfrac{1}{b^2}) = 4(1)$
$$\frac{36}{a^2} + \frac{4}{b^2} = 4 \ \text{...............(3)}$$
$$\frac{4}{a^2} + \frac{4}{b^2} = 1 \ \text{...................(1)}$$

Equation (3) – (1): $\dfrac{32}{a^2} \qquad = 3$

$$a^2 = \frac{32}{3}$$

Substitute $\frac{32}{3}$ for a^2 in equation (2). This gives:

$$\frac{9}{a^2} + \frac{1}{b^2} = 1 \dots\dots\dots\dots(2)$$

$$\frac{9}{\frac{32}{3}} + \frac{1}{b^2} = 1$$

$$\frac{9(3)}{32} + \frac{1}{b^2} = 1$$

$$\frac{1}{b^2} = 1 - \frac{27}{32}$$

$$= \frac{32 - 27}{32}$$

$$\frac{1}{b^2} = \frac{5}{32}$$

$$b^2 = \frac{32}{5}$$

Hence, the equation of the ellipse is:

$$\frac{x^2}{a^2} + \frac{y^2}{b^2} = 1$$

$$\frac{x^2}{\frac{32}{3}} + \frac{y^2}{\frac{32}{5}} = 1$$

$$\frac{3x^2}{32} + \frac{5y^2}{32} = 1$$

10. Find the standard form equation of an ellipse that has vertices (−4, −8) and (−4, 2) and foci (−4, −7) and (−4, 1).

Solution

Since the x-coordinates of the vertices and foci are the same, it means that the major axis is parallel to the y-axis.

The center, (x_1, y_1) of the ellipse is the midpoint of the vertices. The midpoint of the vertices (−4, −8) and (−4, 2) is obtained as follows:

$$\left(\frac{-4 - 4}{2}, \frac{-8 + 2}{2}\right)$$

$$= (-4, -3) \quad \text{(This is } (x_1, y_1))$$

The distance between this midpoint, (−4, −3), and any of the vertices, say (−4, 2) gives 'a'. This is simply obtained by subtracting the lower value of the y coordinates of the two points from the higher value of the y coordinates of the points. This is because the two points have equal x coordinate. Therefore:

$$a = 2 - (-3)$$

$$a = 5$$

So, $a^2 = 25$

Next, we find c. The distance between the center and any of the the foci gives c. Hence, the distance from the center (–4, –3) to the focus (–4, 1) gives c as follows:

c = 1 – (–3)   (The lower y coordinate subtracted from the higher y coordinate)

c = 4

$c^2 = 16$

Recall that: $a^2 = b^2 + c^2$

Hence, $b^2 = a^2 - c^2$

$\quad\quad = 25 - 16$

$b^2 = 9$

Since the major axis is parallel to the y-axis, then the equation of the ellipse will take the form:

$$\frac{(x - x_1)^2}{b^2} + \frac{(y - y_1)^2}{a^2} = 1$$

$$\frac{(x-(-4))^2}{9} + \frac{(y-(-3))^2}{25} = 1 \quad\quad (\text{ Note that } (x_1, y_1) = (-4, -3))$$

$$\frac{(x + 4)^2}{9} + \frac{(y + 3)^2}{25} = 1$$

11. What is the equation of an ellipse with centre (–5, –2), focus (–9, –2) and vertex (0, –2).

Solution

Since the y-coordinates of the vertex and focus are the same, it means that the major axis is parallel to the x-axis.

The center, (x_1, y_1) of the ellipse is (–5, –2).

The distance between the center, (–5, –2), and the vertex, (0, –2) gives 'a'. This is simply obtained by subtracting the lower value of the x coordinates of the two points from the higher value of the x coordinates of the points. This is because the two points have equal y coordinate. Therefore:

a = 0 – (–5)

a = 5

So, $a^2 = 25$

Next, we find c. The distance between the center and the focus gives c. Hence, the distance from the center (–5, –2) to the focus (–9, –2) gives c as follows:

c = –5 – (–9)   (The lower x coordinate subtracted from the higher x coordinate)

c = –5 + 9

c = 4

$c^2 = 16$

Recall that: $a^2 = b^2 + c^2$

Hence, $b^2 = a^2 - c^2$

$$= 25 - 16$$
$$b^2 = 9$$

Since the major axis is parallel to the x-axis, then the equation of the ellipse will take the form:

$$\frac{(x - x_1)^2}{a^2} + \frac{(y - y_1)^2}{b^2} = 1$$

$$\frac{(x-(-5))^2}{9} + \frac{(y-(-2))^2}{25} = 1 \quad \text{(Note that } (x_1, y_1) = (-5, -2))$$

$$\frac{(x + 5)^2}{9} + \frac{(y + 2)^2}{25} = 1$$

12. The equation of an ellipse is given by $2x^2 + 5y^2 + 8x + 10y + 3 = 0$

Express this equation in the canonical/standard form and determine:

(a) the center of the ellipse

(b) the four vertices of the ellipse

(c) the two foci of the ellipse

<u>Solution</u>

(a) Let us complete the square of the terms in x as follows:

$$2x^2 + 5y^2 + 8x + 10y + 3 = 0$$

$$2x^2 + 8x = -5y^2 - 10y - 3 \quad \text{(Leave only the terms in } x \text{ on the left hand side)}$$

Divide each term by 2 to make the coefficient of x^2 to be 1. This gives:

$$x^2 + 4x = -\frac{5}{2}y^2 - 5y - \frac{3}{2}$$

Divide the coefficient of x by 2, square the value obtained and add it to both sides of the equation. The coefficient of x is 4. Then $\frac{4}{2} = 2$. Squaring it gives 2^2. We now add 2^2 to both sides as follows:

$$x^2 + 4x + 2^2 = -\frac{5}{2}y^2 - 5y - \frac{3}{2} + 2^2$$

Take the square root of the left hand side by taking the terms that are squared, and enclosing them in a bracket. Then square the bracket. This gives:

$$(x + 2)^2 = -\frac{5}{2}y^2 - 5y - \frac{3}{2} + 4$$

$$(x + 2)^2 = -\frac{5}{2}y^2 - 5y + \frac{5}{2}$$

The next step is to complete the square of the terms in y by following similar steps.

$$(x + 2)^2 = -\frac{5}{2}y^2 - 5y + \frac{5}{2}$$

$$\frac{5}{2}y^2 + 5y = \frac{5}{2} - (x + 2)^2 \quad \text{(Leave only the terms in y on the left hand side)}$$

Divide each term by $\frac{5}{2}$ to make the coefficient of y^2 to be 1. This gives:

$$y^2 + 2y = 1 - \frac{2}{5}(x+2)^2$$

Divide the coefficient of y by 2, square the value obtained and add it to both sides of the equation. The coefficient of y is 2. Then $\frac{2}{2} = 1$. Squaring it gives 1^2. We now add 1^2 to both sides as follows:

$$y^2 + 2y + 1^2 = 1 - \frac{2}{5}(x+2)^2 + 1^2$$

Take the square root of the left hand side by taking the terms that are squared, and enclosing them in a bracket. Then square the bracket. This gives:

$$(y+1)^2 = 1 - \frac{2}{5}(x+2)^2 + 1$$

$$(y+1)^2 + \frac{2}{5}(x+2)^2 = 2 \qquad \text{(Note that 1 + 1 gives 2 obtained in this step)}$$

Divide each term by 2 to make the right hand side to be 1. This gives:

$$\frac{(y+1)^2}{2} + \frac{(x+2)^2}{5} = 1$$

Or, $\quad \dfrac{(x+2)^2}{5} + \dfrac{(y+1)^2}{2} = 1$

Comparing this equation with the standard equation, $\dfrac{(x-x_1)^2}{a^2} + \dfrac{(y-y_1)^2}{b^2} = 1$, shows that:

$$-x_1 = 2$$
$$x_1 = -2$$

And, $\quad -y_1 = 1$
$$y_1 = -1$$

Therefore, the coordinates of the center i.e. (x_1, y_1) is $(-2, -1)$

(b) Equation: $\quad \dfrac{(x+2)^2}{5} + \dfrac{(y+1)^2}{2} = 1$

Standard form: $\quad \dfrac{(x-x_1)^2}{a^2} + \dfrac{(y-y_1)^2}{b^2} = 1$

By comparison, $a^2 = 5$ while the value of b^2 is 2.

Therefore, $a = \sqrt{5}$ and $b = \sqrt{2}$

Hence, the x-axis is the major axis (from a). The vertices on the major axis are given by:

$\quad V_1 = (a + x_1, \ y_1)$ and $V_2 = (-a + x_1, \ y_1)$. This gives:

$\quad V_1 = (\sqrt{5} - 2, \ -1)$ and $V_2 = (-\sqrt{5} - 2, \ -1)$

The vertices on the minor axis (y-axis) are given by:

$\quad V_3 = (x_1, \ b + y_1)$ and $V_4 = (x_1, \ -b + y_1)$

Therefore, $V_3 = (-2, \ \sqrt{2} - 1)$ and $V_4 = (-2, \ -\sqrt{2} - 1)$

(c) Recall that: $a^2 = b^2 + c^2$

Therefore, $c^2 = a^2 - b^2$

$$c^2 = 5 - 2$$
$$c^2 = 3$$
$$c = \sqrt{3}$$

The foci are given by:

$$F_1 = (c + x_1, \ y_1) \text{ and } F_2 = (-c + x_1, \ y_1)$$

This gives: $F_1 = (\sqrt{3} - 2, \ -1)$ and $F_2 = (-\sqrt{3} - 2, \ -1)$

13. The equation of an ellipse is $9x^2 + 4y^2 - 54x + 16y + 61 = 0$. Write this equation in the canonical form and find:

(a) the center

(b) the vertices

(c) the foci of the ellipse.

Solution

(a) Let us complete the square of the terms in x as follows:

$$9x^2 + 4y^2 - 54x + 16y + 61 = 0$$
$$9x^2 - 54x = -4y^2 - 16y - 61 \quad \text{(Leave only the terms in } x \text{ on the left hand side)}$$

Divide each term by 9 to make the coefficient of x^2 to be 1. This gives:

$$x^2 - 6x = -\frac{4}{9}y^2 - \frac{16}{9}y - \frac{61}{9}$$

Divide the coefficient of x by 2, square the value obtained and add it to both sides of the equation. The coefficient of x is -6. Then $\frac{-6}{2} = -3$. Squaring it gives $(-3)^2$. We now add $(-3)^2$ to both sides as follows:

$$x^2 - 6x + (-3)^2 = -\frac{4}{9}y^2 - \frac{16}{9}y - \frac{61}{9} + (-3)^2$$

Take the square root of the left hand side by taking the terms that are squared, and enclosing them in a bracket. Then square the bracket. This gives:

$$(x - 3)^2 = -\frac{4}{9}y^2 - \frac{16}{9}y - \frac{61}{9} + 9$$
$$(x - 3)^2 = -\frac{4}{9}y^2 - \frac{16}{9}y + \frac{20}{9} \quad \text{(Note that } -\frac{61}{9} + 9 = \frac{20}{9}\text{)}$$

The next step is to complete the square of the terms in y as follows.

$$(x - 3)^2 = -\frac{4}{9}y^2 - \frac{16}{9}y + \frac{20}{9}$$
$$\frac{4}{9}y^2 + \frac{16}{9}y = \frac{20}{9} - (x - 3)^2 \quad \text{(Leave only the terms in y on the left hand side)}$$

Divide each term by $\frac{4}{9}$ to make the coefficient of y^2 to be 1. This gives:

$$y^2 + 4y = 5 - \frac{9}{4}(x - 3)^2$$

Divide the coefficient of y by 2, square the value obtained and add it to both sides of the equation. The coefficient of y is 4. Then $\frac{4}{2} = 2$. Squaring it gives 2^2. We now add 2^2 to both sides as follows:

$$y^2 + 4y + 2^2 = 5 - \frac{9}{4}(x-3)^2 + 2^2$$

Take the square root of the left hand side by taking the terms that are squared, and enclosing them in a bracket. Then square the bracket. This gives:

$$(y + 2)^2 = 5 - \frac{9}{4}(x-3)^2 + 4$$

$$(y + 2)^2 + \frac{9}{4}(x-3)^2 = 9 \qquad \text{(Note that 5 + 4 gives 9 obtained in this step)}$$

Divide each term by 9 to make the right hand side to be 1. This gives:

$$\frac{(y+2)^2}{9} + \frac{(x-3)^2}{4} = 1$$

Or, $\quad \dfrac{(x-3)^2}{4} + \dfrac{(y+2)^2}{9} = 1$

Comparing this equation with the standard equation, $\dfrac{(x-x_1)^2}{b^2} + \dfrac{(y-y_1)^2}{a^2} = 1$, shows that:

$$-x_1 = -3$$
$$x_1 = 3$$

And, $\quad -y_1 = 2$
$$y_1 = -2$$

Therefore, the coordinates of the center i.e. (x_1, y_1) is (3, −2)

(b) Equation: $\quad \dfrac{(x-3)^2}{4} + \dfrac{(y+2)^2}{9} = 1$

Standard form: $\quad \dfrac{(x-x_1)^2}{b^2} + \dfrac{(y-y_1)^2}{a^2} = 1$

By comparison, $a^2 = 9$ while the value of b^2 is 4.

Therefore, $a = \sqrt{9}$
$$a = 3$$

and $\quad b = \sqrt{4}$
$$b = 2$$

Hence, the y-axis is the major axis (from a). The vertices on the major axis are given by:

$\quad V_1 = (x_1, \ a + y_1)$ and $V_2 = (x_1, \ -a + y_1)$. This gives:
$\quad V_1 = (3, \ 3-2)$ and $V_2 = (3, \ -3-2)$
$\quad V_1 = (3, 1)$ and $V_2 = (3, -5)$

(c) Recall that: $a^2 = b^2 + c^2$

Therefore, $c^2 = a^2 - b^2$
$$c^2 = 9 - 4$$
$$c^2 = 5$$

$$c = \sqrt{5}$$

The foci are given by:

$$F_1 = (x_1, \ c + y_1) \text{ and } F_2 = (x_1, \ -c + y_1)$$

This gives: $F_1 = (3, \ \sqrt{5} - 2)$ and $F_2 = (3, \ -\sqrt{5} - 2)$

14. The equation of an ellipse is $4x^2 + 9y^2 = 36$. Find the equation of the tangent to the ellipse at the point $(0, -2)$.

Solution

$$4x^2 + 9y^2 = 36$$

Dividing each term by 36 gives:

$$\frac{x^2}{9} + \frac{y^2}{4} = 1$$

Hence, $a^2 = 9$ and $b^2 = 4$

The equation of the tangent to the ellipse is given by:

$$\frac{xx_1}{a^2} + \frac{yy_1}{b^2} = 1$$

Or, $b^2 xx_1 + a^2 yy_1 = a^2 b^2$

Substitute $(0, -2)$ for (x_1, y_1), 9 for a^2 and 4 for b^2, in the equation above. This gives:

$$b^2 xx_1 + a^2 yy_1 = a^2 b^2$$
$$4(x)(0) + 9(y)(-2) = 9(4)$$
$$0 - 18y = 36$$
$$y = -2$$

Therefore, the equation of the tangent is $y = -2$

15. The equation of an ellipse is given by $2x^2 + 9y^2 = 18$. Find the equation of the normal to the ellipse at the point $(1, \frac{4}{3})$.

Solution

$$2x^2 + 9y^2 = 18$$

Dividing each term by 18 gives:

$$\frac{x^2}{9} + \frac{y^2}{2} = 1$$

Hence, $a^2 = 9$ and $b^2 = 2$

The equation of the normal to the ellipse is given by:

$$a^2 xy_1 - b^2 x_1 y = (a^2 - b^2) x_1 y_1$$

Substitute $(1, \frac{4}{3})$ for (x_1, y_1), 9 for a^2 and 2 for b^2, in the equation above. This gives:

$$a^2 xy_1 - b^2 x_1 y = (a^2 - b^2) x_1 y_1$$

$$9(x)(\tfrac{4}{3}) - 2(1)(y) = (9-2)(1)(\tfrac{4}{3})$$

$$12x - 2y = 7(\tfrac{4}{3})$$

$$12x - 2y = \frac{28}{3}$$

Multiply each term by 3 to clear out the fraction. This gives:

$$36x - 6y = 28$$

$$18x - 3y - 14 = 0 \quad \text{(After each term is divided by 2)}$$

Therefore the equation of the normal to the ellipse is $18x - 3y - 14 = 0$

Exercise 6

1 Find the four vertices and the foci of an ellipse whose equation is: $\dfrac{x^2}{16} + \dfrac{y^2}{36} = 1$

2. Find the vertices and the foci of the ellipse whose equation is: $\dfrac{x^2}{100} + \dfrac{y^2}{64} = 1$

3. Find the values of a, b and c in the equation of the ellipse: $25x^2 + 4y^2 = 100$

4. Find the length of the major and minor axes of the ellipse $\dfrac{x^2}{20} + \dfrac{y^2}{45} = 1$

5. Find the equation of an ellipse with vertices $(0, \pm10)$ and foci $(0, \pm8)$.

6. Find the equation of an ellipse with vertices $(\pm17, 0)$ and foci $(\pm15, 0)$.

7. The length of the semi-major and semi-minor axes of an ellipse are 5 and 3 units respectively. Find the:

(a) eccentricity

(b) length of the latus rectum

(c) the area of the ellipse

8. An ellipse has foci $(\pm12, 0)$ and eccentricity, e $= \dfrac{12}{13}$. Find the equation of the ellipse.

9. Find the equation of the ellipse having major axis along the x-axis and passing through the points (6, 4) and (−8,3).

10. Find the standard form equation of an ellipse that has vertices (−2, −1) and (−2, 6) and foci (−2, 1) and (−2, 4).

11. What is the equation of an ellipse with centre (2, 3), focus (−5, 3) and vertex (−8, 3).

12. The equation of an ellipse is given by $9x^2 + 4y^2 - 90x + 24y + 225 = 0$

Express this equation in the canonical/standard form and determine:

(a) the center of the ellipse

(b) the four vertices of the ellipse

(c) the two foci of the ellipse

13. The equation of an ellipse is $x^2 + 4y^2 - 2x - 16y + 1 = 0$. Write this equation in the canonical form and find:

(a) the center

(b) the vertices

(c) the foci of the ellipse.

14. The equation of an ellipse is $16x^2 + 25y^2 = 400$. Find the equation of the tangent to the ellipse at the point $(-5, 0)$.

15. The equation of an ellipse is given by $x^2 + 4y^2 = 4$. Find the equation of the normal to the ellipse at the point $(0, 1)$.

CHAPTER 7
THE HYPERBOLA

What is a Hyperbola

A hyperbola is the locus of a point moving in a plane such that its distance from two fixed points has a constant difference. The two fixed points are the foci of the hyperbola.

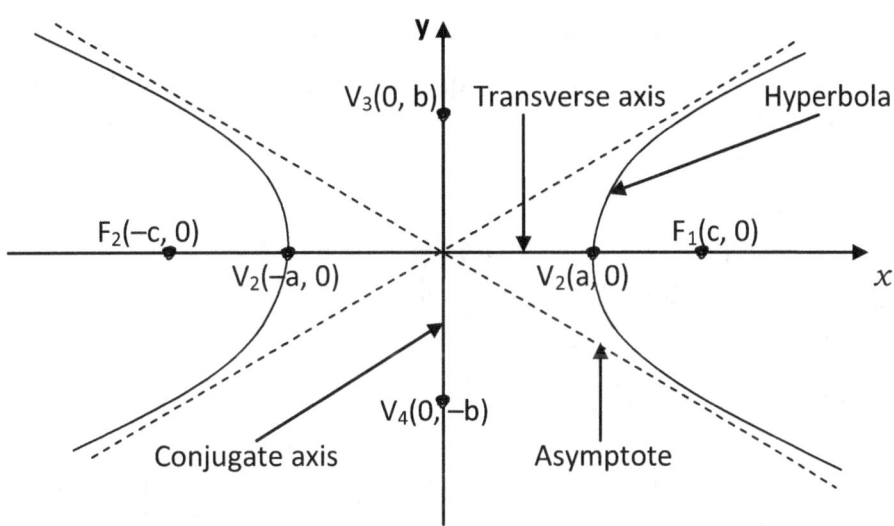

Equation of a Hyperbola with Center at the Origin, (0, 0)

1. The standard form of the equation of a hyperbola whose center is at the origin (0, 0) and transverse axis on the x-axis is given by:

$$\frac{x^2}{a^2} - \frac{y^2}{b^2} = 1$$ (Letter a always comes first on the left as the positive part of the equation)

where a = half of the length of the transverse axis

b = half of the length of the conjugate axis

$V_1 = (a, 0)$, and $V_2 = (-a, 0)$ are the vertices

$V_3 = (0, b)$ and $V_4 = (0, -b)$ are the co-vertices

$F_1 = (c, 0)$ and $F_2 = (-c, 0)$ are the foci

c = distance between the focus and the center, and $c^2 = a^2 + b^2$

The equations of the asymptotes are $y = \frac{b}{a}x$ and $y = -\frac{b}{a}x$

2. The standard form of the equation of a hyperbola whose center is at the origin (0, 0) and transverse axis on the y-axis is given by:

$$\frac{y^2}{a^2} - \frac{x^2}{b^2} = 1$$ (Letter a always comes first on the left as the positive part of the equation)

where a = half of the length of the transverse axis

b = half of the length of the conjugate axis

$V_1 = (0, a)$, and $V_2 = (0, -a)$ are the vertices

$V_3 = (b, 0)$ and $V_4 = (-b, 0)$ are the co-vertices

$F_1 = (0, c)$ and $F_2 = (0, -c)$ are the foci

c = distance between the focus and the center, and $c^2 = a^2 + b^2$

The equations of the asymptotes are $y = \dfrac{a}{b}x$ and $y = -\dfrac{a}{b}x$

Equation of a Hyperbola when the Center not at the Origin

1. The standard form of the equation of a hyperbola with center (x_1, y_1) and transverse axis parallel to the x-axis is given by:
$$\frac{(x - x_1)^2}{a^2} - \frac{(y - y_1)^2}{b^2} = 1$$
where $V_1 = (a + x_1, y_1)$, and $V_2 = (-a + x_1, y_1)$ are the vertices

$V_3 = (x_1, b + y_1)$ and $V_4 = (x_1, -b + y_1)$ are the co-vertices

$F_1 = (c + x_1, y_1)$ and $F_2 = (-c + x_1, y_1)$ are the foci

The equations of the asymptotes are $y = \dfrac{b}{a}(x - x_1) + y_1$ and $y = -\dfrac{b}{a}(x - x_1) + y_1$

2. The standard form of the equation of a hyperbola with center (x_1, y_1) and transverse axis parallel to the y-axis is given by:
$$\frac{(y - y_1)^2}{a^2} - \frac{(x - x_1)^2}{b^2} = 1$$
where $V_1 = (x_1, a + y_1)$, and $V_2 = (x_1, -a + y_1)$ are the vertices

$V_3 = (b + x_1, y_1)$ and $V_4 = (-b + x_1, y_1)$ are the co-vertices

$F_1 = (x_1, c + y_1)$ and $F_2 = (x_1, -c + y_1)$ are the foci

The equations of the asymptotes are $y = \dfrac{a}{b}(x - x_1) + y_1$ and $y = -\dfrac{a}{b}(x - x_1) + y_1$

Latus Rectum of a Hyperbola

The length of the latus rectum of a hyperbola is given by:
$$L = \frac{2b^2}{a}$$

Eccentricity of a hyperbola

The eccentricity of a hyperbola is given by:
$$e = \sqrt{1 + \frac{b^2}{a^2}} \quad \text{or} \quad e = \frac{c}{a}$$

Equation of the Tangent to a Hyperbola

The equation of the tangent to the hyperbola $\dfrac{x^2}{a^2} - \dfrac{y^2}{b^2} = 1$ at the point (x_1, y_1) is given by:

$$\frac{xx_1}{a^2} - \frac{yy_1}{b^2} = 1$$

Equation of the Normal to a Hyperbola

The equation of the normal to the hyperbola $\dfrac{x^2}{a^2} - \dfrac{y^2}{b^2} = 1$ at the point (x_1, y_1) is given by:

$$\frac{a^2 x}{x_1} - \frac{b^2 y}{y_1} = a^2 + b^2$$

Examples

1. The equation of a hyperbola is given by $9x^2 - 4y^2 = 36$. Find the vertices and foci of the hyperbola.

Solution

$$9x^2 - 4y^2 = 36$$

Dividing each term by 36 gives:

$$\frac{x^2}{4} - \frac{y^2}{9} = 1 \qquad \text{(Center is the origin)}$$

Comparing this equation with the standard form equation, $\dfrac{x^2}{a^2} - \dfrac{y^2}{b^2} = 1$, shows that:

$$a^2 = 4 \text{ and } b^2 = 9$$

Hence $a = 2$ and $b = 3$ (After taking the square roots of a^2 and b^2)

Since a^2 is under x^2 in the equation, it shows that the transverse axis is the x-axis. This means that the vertices and foci lie on the x-axis.

Hence, $V_1 = (a, 0)$ and $V_2 = (-a, 0)$

Therefore, $V_1 = (2, 0)$ and $V_2 = (-2, 0)$

Recall that: $c^2 = a^2 + b^2$

$$c^2 = 4 + 9$$
$$c = \sqrt{13}$$

Hence, $F_1 = (c, 0)$ and $F_2 = (-c, 0)$

Therefore, $F_1 = (\sqrt{13}, 0)$ and $F_2 = (-\sqrt{13}, 0)$

2. The equation of a hyperbola is given by, $16x^2 - 9y^2 = -144$. Find:

(a) the length of its transverse axis

(b) the length of its conjugate axis

(c) its eccentricity

(d) its foci

(e) its vertices

(f) the length of its latus rectum

(g) the equation of its asymptotes

Solution

$$16x^2 - 9y^2 = -144$$

Dividing each term by −144 gives:

$$-\frac{x^2}{9} + \frac{y^2}{16} = 1 \quad \text{(Center is the origin)}$$

Or, $\quad \dfrac{y^2}{16} - \dfrac{x^2}{9} = 1$

Comparing this equation with the standard form equation, $\dfrac{y^2}{a^2} - \dfrac{x^2}{b^2} = 1$, shows that:

$$a^2 = 16 \text{ and } b^2 = 9$$

Hence a = 4 and b = 3 (After taking the square roots of a^2 and b^2)

Recall that: $c^2 = a^2 + b^2$

$$c^2 = 16 + 9$$

$$c = \sqrt{25}$$

$$c = 5$$

(a) Length of the transverse axis = 2a (Recall that a = half the length of the transverse axis)

$$= 2(4)$$

$$= 8 \text{ units}$$

(b) Length of the conjugate axis = 2b (Recall that b = half the length of the conjugate axis)

$$= 2(3)$$

$$= 6 \text{ units}$$

(c) Eccentricity, $e = \dfrac{c}{a}$

$$= \frac{5}{4}$$

$$c = 1.25$$

(d) $\quad \dfrac{y^2}{a^2} - \dfrac{x^2}{b^2} = 1$

Since a^2 is under y^2 in the equation, it shows that the transverse axis is the y-axis. This means that the vertices and foci lie on the y-axis.

Hence, $F_1 = (0, c)$ and $F_2 = (0, -c)$

Therefore, $F_1 = (0, 5)$ and $F_2 = (0, -5)$

(e) Since the vertices are also on the y-axis, they are given by:

$$V_1 = (0, a) \text{ and } V_2 = (0, -a)$$

Hence, $V_1 = (0, 4)$ and $V_2 = (0, -4)$

(f) $L = \dfrac{2b^2}{a}$

$= \dfrac{2(3)^2}{4}$

$= \dfrac{2(9)}{4}$

$= 4.5$ units

(g) The equation of the asymptotes are given by:

$y = \dfrac{a}{b}x$

$y = \dfrac{4}{3}x$

Hence, $3y = 4x$

And, $y = -\dfrac{a}{b}x$

$y = -\dfrac{4}{3}x$

Hence, $3y = -4x$

3. The length of the latus rectum of a hyperbola is half of its transverse axis. Find the eccentricity of the hyperbola.

Solution

Half of the transverse axis of a hyperbola is a.

The length of the latus rectum is given by: $L = \dfrac{2b^2}{a}$

From the question: The length of the latus rectum = Half of the transverse axis

Hence, $\dfrac{2b^2}{a} = a$

$a^2 = 2b^2$

Recall that: Eccentricity, $e = \sqrt{1 + \dfrac{b^2}{a^2}}$

$e = \sqrt{1 + \dfrac{b^2}{2b^2}}$ (After putting $a^2 = 2b^2$)

$= \sqrt{1 + \dfrac{1}{2}}$ (b^2 cancels out)

$e = \sqrt{\dfrac{3}{2}}$

$e = 1.225$

4. Find the equation of a hyperbola with vertices $(0, \pm 8)$ and foci $(0, \pm 10)$.

Solution

Since the x coordinates of the vertices and foci are the same (i.e. 0) it shows that the transverse axis is on the y-axis. Hence, $V_1 = (0, a)$ and $V_2 = (0, -a)$. Also, $F_1 = (0, c)$ and $F_2 = (0, -c)$.

Comparing these points with the coordinates given in the question shows that:

$V_1 = (0, 8)$, $V_2 = (0, -8)$, $F_1 = (0, 10)$ and $F_2 = (0, -10)$

Hence, a = 8 and c = 10.

Recall that: $c^2 = a^2 + b^2$

Hence, $b^2 = c^2 - a^2$

$= 10^2 - 8^2$

$= 100 - 64$

$b^2 = 36$

Note that $a^2 = 8^2$

Hence, $a^2 = 64$

Therefore, the equation of the hyperbola is:

$$\frac{y^2}{a^2} - \frac{x^2}{b^2} = 1 \qquad \text{(Note that a is on the y-axis, the transverse axis)}$$

Therefore the equation of the hyperbola is $\dfrac{y^2}{64} - \dfrac{x^2}{36} = 1$

5. Find the equation of a hyperbola with vertices $(\pm 5, 0)$ and foci $(\pm 13, 0)$.

Solution

Since the y coordinates of the vertices and foci are the same (i.e. 0) it shows that the transverse axis is on the x-axis. Hence, the x coordinates of the vertices shows that a = 5 while the x coordinates of the foci shows that c = 13.

Recall that: $c^2 = a^2 + b^2$

$b^2 = c^2 - a^2$

$= 13^2 - 5^2$

$= 169 - 25$

$b^2 = 144$

Note that $a^2 = 5^2$

Hence, $a^2 = 25$

Therefore, the equation of the hyperbola is:

$$\frac{x^2}{a^2} - \frac{y^2}{b^2} = 1 \qquad \text{(Note that a is on the x-axis, the transverse axis)}$$

Therefore the equation of the hyperbola is $\dfrac{x^2}{25} - \dfrac{y^2}{144} = 1$

6. The values of a and b of a hyperbola are 4 and 9 units respectively. Find the:

(a) eccentricity

(b) length of the latus rectum of the hyperbola

Solution

(a) $a = 4$

$b = 9$

Eccentricity, $e = \sqrt{1 + \dfrac{b^2}{a^2}}$

$= \sqrt{1 + \dfrac{9^2}{4^2}}$

$= \sqrt{1 + \dfrac{81}{16}}$

$= \sqrt{\dfrac{97}{16}}$

$e = 2.46$

(b) The length of the latus rectum is given by:

$L = \dfrac{2b^2}{a}$

$= \dfrac{2(9)^2}{4}$

$= \dfrac{2(81)}{4}$

$= \dfrac{81}{2}$

$L = 40.5$ units

7. A hyperbola has foci $(\pm 6, 0)$ and eccentricity, $e = 1.25$. Find the equation of the hyperbola.

Solution

The foci is $(\pm 6, 0)$. This shows that the transverse axis is on the x-axis since $y = 0$ on the foci. It also shows that the center of the hyperbola is the origin $(0, 0)$.

Recall that foci $(\pm 6, 0) = (\pm c, 0)$. Therefore, $c = 6$.

Eccentricity, $e = \dfrac{c}{a}$

$1.25 = \dfrac{6}{a}$

$1\dfrac{1}{4} = \dfrac{6}{a}$ (Note that $1.25 = 1\dfrac{1}{4}$ when expressed in fraction)

$\dfrac{5}{4} = \dfrac{6}{a}$

$5a = 6(4)$

127

$$a = \frac{24}{5}$$

Recall also that:

$$c^2 = a^2 + b^2$$

$$b^2 = c^2 - a^2$$

$$= 6^2 - (\frac{24}{5})^2$$

$$= 36 - \frac{576}{25}$$

$$= \frac{900 - 576}{25}$$

$$b^2 = \frac{324}{25}$$

Note that $a^2 = (\frac{24}{5})^2$

$$a^2 = \frac{576}{25}$$

Since the x-axis is the transverse axis, the equation will take the form:

$$\frac{x^2}{a^2} - \frac{y^2}{b^2} = 1$$

This gives: $\dfrac{x^2}{\frac{576}{25}} + \dfrac{y^2}{\frac{324}{25}} = 1$

This simplifies to: $\dfrac{25x^2}{576} - \dfrac{25y^2}{324} = 1$

8. Find the equation of the hyperbola having transverse axis along the x-axis and passing through the points (3, 0) and (–6, 3).

Solution

Since the major axis is on the x-axis, then the equation takes the form:

$$\frac{x^2}{a^2} - \frac{y^2}{b^2} = 1 \quad \text{('a' goes under the transverse axis, i.e } x\text{-axis)}$$

Substituting the first point (3, 0) into the equation above gives:

$$\frac{3^2}{a^2} - \frac{(0)^2}{b^2} = 1$$

$$\frac{9}{a^2} - 0 = 1$$

$$\frac{9}{a^2} = 1 \quad \text{................(1)}$$

Similarly, substituting the second point (–6, 3) gives:

$$\frac{(-6)^2}{a^2} - \frac{(3)^2}{b^2} = 1$$

$$\frac{36}{a^2} - \frac{9}{b^2} = 1 \quad \text{...................(2)}$$

From equation (1) $a^2 = 9$ ....(3)

Substitute 9 for a^2 in equation (2). This is done as follows:

$$\frac{36}{a^2} - \frac{9}{b^2} = 1 \quad ....(2)$$

$$\frac{36}{9} - \frac{9}{b^2} = 1$$

$$4 - \frac{9}{b^2} = 1$$

$$4 - 1 = \frac{9}{b^2}$$

$$3 = \frac{9}{b^2}$$

$$b^2 = \frac{9}{3}$$

$$b^2 = 3$$

Note that $a^2 = 9$ (From equation (3) above)

Hence, the equation of the hyperbola is:

$$\frac{x^2}{a^2} - \frac{y^2}{b^2} = 1$$

$$\frac{x^2}{9} - \frac{y^2}{3} = 1$$

9. Find the standard form equation of a hyperbola that has vertices (−2, 3) and (−2, 1) and foci (−2, 5) and (−2, −1).

Solution

Since the x-coordinates of the vertices and foci are the same, it means that the transverse axis is parallel to the y-axis.

The center, (x_1, y_1) of the hyperbola is the midpoint of the vertices. The midpoint of the vertices (−2, 3) and (−2, 1) is obtained as follows:

$$\left(\frac{-2 - 2}{2}, \ \frac{3 + 1}{2} \right)$$

$= (-2, 2)$ (This is (x_1, y_1), i.e. the center)

The distance between this center, (−2, 2), and any of the vertices, say (−2, 1) gives 'a'. This is simply obtained by subtracting the lower value of the y coordinates of the two points from the higher value of the y coordinates of the points. This is because the two points have equal x coordinate. Therefore:

$$a = 2 - 1$$

$$a = 1$$

So, $a^2 = 1$

Next, we find c. The distance between the center and any of the foci gives c. Hence, the distance from the center (−2, 2) to the focus (−2, −1) gives c as follows:

c = 2 − (−1)   (The lower y coordinate subtracted from the higher y coordinate)

c = 3

$c^2 = 9$

Recall that: $c^2 = a^2 + b^2$

Hence, $b^2 = c^2 - a^2$

$\qquad = 9 - 1$

$\qquad b^2 = 8$

Since the transverse axis is parallel to the y-axis, then the equation of the hyperbola will take the form:

$$\frac{(y - y_1)^2}{a^2} - \frac{(x - x_1)^2}{b^2} = 1$$

$$\frac{(y - 2)^2}{1} - \frac{(x - (-2))^2}{8} = 1 \quad (\text{Note that center } (x_1, y_1) = (-2, 2))$$

$$\frac{(y - 2)^2}{1} - \frac{(x + 2)^2}{8} = 1$$

10. What is the equation of a hyperbola with centre at (3, −1), focus at (9, −1) and vertex at (5, −1).

Solution

Since the y-coordinates of the vertex and focus are the same, it means that the transverse axis is parallel to the x-axis.

The center, (x_1, y_1) of the hyperbola is (3, −1).

The distance between the center, (3, −1), and the vertex, (5, −1) gives 'a'. This is simply obtained by subtracting the lower value of the x coordinates of the two points from the higher value of the x coordinates of the points. This is because the two points have equal y coordinate. Therefore:

$\qquad a = 5 - 3$

$\qquad a = 2$

So, $a^2 = 4$

Next, we find c. The distance between the center and the focus gives c. Hence, the distance from the center (3, −1) to the focus (9, −1) gives c as follows:

$\qquad c = 9 - 3$ (The lower x coordinate subtracted from the higher x coordinate)

$\qquad c = 6$

$\qquad c^2 = 36$

Recall that: $c^2 = a^2 + b^2$

Hence, $b^2 = c^2 - a^2$

$\qquad = 36 - 4$

$\qquad b^2 = 32$

Since the transverse axis is parallel to the x-axis, then the equation of the hyperbola will take the form:

$$\frac{(x - x_1)^2}{a^2} - \frac{(y - y_1)^2}{b^2} = 1$$

$$\frac{(x-3)^2}{4} - \frac{(y-(-1))^2}{32} = 1 \quad \text{(Note that } (x_1, y_1) = (3, -1))$$

$$\frac{(x-3)^2}{4} - \frac{(y+1)^2}{32} = 1$$

11. The equation of a hyperbola is given by $4x^2 - 9y^2 - 16x - 90y - 245 = 0$

Express this equation in the canonical/standard form and determine:

(a) the center of the hyperbola

(b) the four vertices of the hyperbola

(c) the two foci of the hyperbola

Solution

(a) Let us complete the square of the terms in x as follows:

$$4x^2 - 9y^2 - 16x - 90y - 245 = 0$$

$$4x^2 - 16x = 9y^2 + 90y + 245 \quad \text{(Leave only the terms in } x \text{ on the left hand side)}$$

Divide each term by 4 to make the coefficient of x^2 to be 1. This gives:

$$x^2 - 4x = \frac{9}{4}y^2 + \frac{45}{2}y + \frac{245}{4}$$

Divide the coefficient of x by 2, square the value obtained and add it to both sides of the equation. The coefficient of x is -4. Then $\frac{-4}{2} = -2$. Squaring it gives -2^2. We now add -2^2 to both sides as follows:

$$x^2 - 4x + (-2)^2 = \frac{9}{4}y^2 + \frac{45}{2}y + \frac{245}{4} + (-2)^2$$

Take the square root of the left hand side by taking the terms that are squared, and enclosing them in a bracket. Then square the bracket. This gives:

$$(x-2)^2 = \frac{9}{4}y^2 + \frac{45}{2}y + \frac{245}{4} + 4$$

$$(x-2)^2 = \frac{9}{4}y^2 + \frac{45}{2}y + \frac{261}{4}$$

The next step is to complete the square of the terms in y by following similar steps.

$$(x-2)^2 = \frac{9}{4}y^2 + \frac{45}{2}y + \frac{261}{4}$$

$$-\frac{9}{4}y^2 - \frac{45}{2}y = \frac{261}{4} - (x-2)^2 \quad \text{(Leave only the terms in y on the left hand side)}$$

Divide each term by $-\frac{9}{4}$ to make the coefficient of y^2 to be 1. This gives:

$$y^2 + 10y = -29 + \frac{4}{9}(x-2)^2$$

Divide the coefficient of y by 2, square the value obtained and add it to both sides of the equation. The coefficient of y is 10. Then $\frac{10}{2}$ = 5. Squaring it gives 5^2. We now add 5^2 to both sides as follows:

$$y^2 + 10y + 5^2 = -29 + \frac{4}{9}(x-2)^2 + 5^2$$

Take the square root of the left hand side by taking the terms that are squared, and enclosing them in a bracket. Then square the bracket. This gives:

$$(y + 5)^2 = -29 + \frac{4}{9}(x-2)^2 + 25$$

$$(y + 5)^2 - \frac{4}{9}(x-2)^2 = -4 \qquad \text{(Note that } -29 + 25 = -4 \text{ as obtained in this step)}$$

Divide each term by –4 to make the right hand side to be 1. This gives:

$$\frac{(x-2)^2}{9} - \frac{(y+5)^2}{4} = 1$$

Comparing this equation with the standard equation, $\frac{(x-x_1)^2}{a^2} - \frac{(y-y_1)^2}{b^2} = 1$, shows that:

$$-x_1 = -2$$
$$x_1 = 2$$

And, $-y_1 = 5$
$$y_1 = -5$$

Therefore, the coordinates of the center i.e. (x_1, y_1) is (2, –5)

(b) Equation: $\frac{(x-2)^2}{9} - \frac{(y+5)^2}{4} = 1$

Standard form: $\frac{(x-x_1)^2}{a^2} - \frac{(y-y_1)^2}{b^2} = 1$

By comparison, $a^2 = 9$ while the value of b^2 is 4.

Therefore, a = 3 and b = 2

Hence, the x-axis is parallel to the transverse axis. The vertices on the transverse axis are given by:

$V_1 = (a + x_1, \ y_1)$ and $V_2 = (-a + x_1, \ y_1)$. This gives:
$V_1 = (3 + 2, \ -5)$ and $V_2 = (-3 + 2, \ -5)$
$V_1 = (5, -5)$ and $V_2 = (-1, -5)$

The vertices on the conjugate axis (parallel to the y-axis) are given by:
$V_3 = (x_1, \ b + y_1)$ and $V_4 = (x_1, \ -b + y_1)$

Therefore, $V_3 = (2, 2 - 5)$ and $V_4 = (2, -2 - 5)$
$V_3 = (2, -3)$ and $V_4 = (2, -7)$

(c) Recall that: $c^2 = a^2 + b^2$
$c^2 = 9 + 4$
$c^2 = 13$

$$c = \sqrt{13}$$

The foci are given by:

$$F_1 = (c + x_1,\ y_1) \text{ and } F_2 = (-c + x_1,\ y_1)$$

This gives: $F_1 = (\sqrt{13} + 2,\ -5)$ and $F_2 = (-\sqrt{13} + 2,\ -5)$

12. The equation of a hyperbola is $16y^2 - 9x^2 + 96y + 72x - 576 = 0$. Write this equation in the canonical form and find:

(a) the center

(b) the vertices

(c) the foci of the hyperbola.

Solution

(a) Let us complete the square of the terms in x as follows:

$$16y^2 - 9x^2 + 96y + 72x - 576 = 0$$
$$-9x^2 + 72x = -16y^2 - 96y + 576$$

Divide each term by -9 to make the coefficient of x^2 to be 1. This gives:

$$x^2 - 8x = \frac{16}{9}y^2 + \frac{32}{3}y - 64$$

Divide the coefficient of x by 2, square the value obtained and add it to both sides of the equation. The coefficient of x is -8. Then $\frac{-8}{2} = -4$. Squaring it gives -4^2. We now add -4^2 to both sides as follows:

$$x^2 - 8x + (-4)^2 = \frac{16}{9}y^2 + \frac{32}{3}y - 64 + (-4)^2$$

Take the square root of the left hand side by taking the terms that are squared, and enclosing them in a bracket. Then square the bracket. This gives:

$$(x - 4)^2 = \frac{16}{9}y^2 + \frac{32}{3}y - 64 + 16$$
$$(x - 4)^2 = \frac{16}{9}y^2 + \frac{32}{3}y - 48$$

The next step is to complete the square of the terms in y.

$$(x - 4)^2 = \frac{16}{9}y^2 + \frac{32}{3}y - 48$$
$$-\frac{16}{9}y^2 - \frac{32}{3}y = -48 - (x - 4)^2$$

Divide each term by $-\frac{16}{9}$ to make the coefficient of y^2 to be 1. This gives:

$$y^2 + 6y = 27 + \frac{9}{16}(x - 4)^2$$

Divide the coefficient of y by 2, square the value obtained and add it to both sides of the equation. The coefficient of y is 6. Then $\frac{6}{2} = 3$. Squaring it gives 3^2. We now add 3^2 to both sides as follows:

133

$$y^2 + 6y + 3^2 = 27 + \frac{9}{16}(x-4)^2 + 3^2$$

Take the square root of the left hand side by taking the terms that are squared, and enclosing them in a bracket. Then square the bracket. This gives:

$$(y+3)^2 = 27 + \frac{9}{16}(x-4)^2 + 9$$

$$(y+3)^2 - \frac{9}{16}(x-4)^2 = 36$$

Divide each term by 36 to make the right hand side to be 1. This gives:

$$\frac{(y+3)^2}{36} - \frac{(x-4)^2}{64} = 1$$

Comparing this equation with the standard equation, $\frac{(y-y_1)^2}{a^2} - \frac{(x-x_1)^2}{b^2} = 1$, shows that:

$$-y_1 = 3$$
$$y_1 = -3$$
And, $-x_1 = -4$
$$x_1 = 4$$

Therefore, the coordinates of the center i.e. (x_1, y_1) is (4, −3)

(b) Equation: $\dfrac{(y+3)^2}{36} - \dfrac{(x-4)^2}{64} = 1$

Standard form: $\dfrac{(y-y_1)^2}{a^2} - \dfrac{(x-x_1)^2}{b^2} = 1$

By comparison, $a^2 = 36$ while the value of b^2 is 64.

Therefore, a = 6 and b = 8

Hence, the y-axis is parallel to the transverse axis. The vertices on the transverse axis are given by:

$V_1 = (x_1, a + y_1)$ and $V_2 = (x_1, -a + y_1)$. This gives:
$V_1 = (4, 6 - 3)$ and $V_2 = (4, -6 - 3)$
$V_1 = (4, 3)$ and $V_2 = (4, -9)$

(c) Recall that: $c^2 = a^2 + b^2$
$$c^2 = 36 + 64$$
$$c^2 = 100$$
$$c = \sqrt{100}$$
$$c = 10$$

The foci are given by:

$F_1 = (x_1, c + y_1)$ and $F_2 = (x_1, -c + y_1)$. This gives:
This gives: $F_1 = (4, 10 - 3)$ and $F_2 = (4, -10 - 3)$
$F_1 = (4, 7)$ and $F_2 = (4, -13)$.

13. The equation of a hyperbola is $25x^2 - 4y^2 = 100$. Find the equation of the tangent to the hyperbola at the point (2, 0).

<u>Solution</u>

$$25x^2 - 4y^2 = 100$$

Dividing each term by 100 gives:

$$\frac{x^2}{4} - \frac{y^2}{25} = 1$$

Hence, $a^2 = 4$ and $b^2 = 25$

The equation of the tangent to the hyperbola is given by:

$$\frac{xx_1}{a^2} - \frac{yy_1}{b^2} = 1$$

Substitute (2, 0) for (x_1, y_1), 4 for a^2 and 25 for b^2, in the equation above. This gives:

$$\frac{xx_1}{a^2} - \frac{yy_1}{b^2} = 1$$

$$\frac{x(2)}{4} - \frac{y(0)}{25} = 1$$

$$\frac{x}{2} - 0 = 1$$

$$x = 2$$

Therefore, the equation of the tangent is $x = 2$

14. The equation of a hyperbola is given by $x^2 - 4y^2 = 64$. Find the equation of the normal to the hyperbola at the point (10, 3).

<u>Solution</u>

$$x^2 - 4y^2 = 64$$

Dividing each term by 64 gives:

$$\frac{x^2}{64} - \frac{y^2}{16} = 1$$

Hence, $a^2 = 64$ and $b^2 = 16$

The equation of the normal to the hyperbola is given by:

$$\frac{a^2 x}{x_1} - \frac{b^2 y}{y_1} = a^2 + b^2$$

Substitute (10, 3) for (x_1, y_1), 64 for a^2 and 16 for b^2, in the equation above. This gives:

$$\frac{a^2 x}{x_1} - \frac{b^2 y}{y_1} = a^2 + b^2$$

$$\frac{64x}{10} - \frac{16y}{3} = 64 + 16$$

$$\frac{64x}{10} - \frac{16y}{3} = 80$$

$$\frac{32x}{5} - \frac{16y}{3} = 80$$

Multiply each term by 15 (LCM of 5 and 3) in order to clear out the fractions. This gives:

$96x - 80y = 1200$

Divide each term by 16 to express it in its lowest term. This gives:

$6x - 5y = 75$

Therefore the equation of the normal to the hyperbola is $6x - 5y - 75 = 0$

Exercise 7

1. The equation of a hyperbola is given by $25x^2 - 4y^2 = 100$. Find the vertices and foci of the hyperbola.

2. The equation of a hyperbola is given by, $36x^2 - 100y^2 = -3600$. Find:

(a) the length of its transverse axis

(b) the length of its conjugate axis

(c) its eccentricity

(d) its foci

(e) its vertices

(f) the length of its latus rectum

(g) the equation of its asymptotes

3. The length of the latus rectum of a hyperbola is one fourth of its transverse axis. Find the eccentricity of the hyperbola.

4. Find the equation of a hyperbola with vertices $(0, \pm 3)$ and foci $(0, \pm 5)$.

5. Find the equation of a hyperbola with vertices $(\pm 8, 0)$ and foci $(\pm 17, 0)$.

6. The values of a and b of a hyperbola are 16 and 25 units respectively. Find the:

(a) eccentricity

(b) length of the latus rectum of the hyperbola

7. A hyperbola has foci $(\pm 5, 0)$ and eccentricity, e = 1.33. Find the equation of the hyperbola.

8. Find the equation of a hyperbola having transverse axis along the x-axis and passing through the points (4, 3) and (2, 1).

9. Find the standard form equation of a hyperbola that has vertices (1, −2) and (1, −6) and foci (1, 0) and (1, −8).

10. What is the equation of a hyperbola whose centre is at (−2, 2), focus at (4, 2) and vertex at (1, 2).

11. The equation of a hyperbola is given by $9x^2 - 4y^2 - 36x - 16y - 16 = 0$

Express this equation in the canonical/standard form and determine:

(a) the center of the hyperbola

(b) the four vertices of the hyperbola

(c) the two foci of the hyperbola

12. The equation of a hyperbola is $25y^2 - 4x^2 - 50y - 16x - 91 = 0$. Write this equation in the canonical form and find:

(a) the center

(b) the vertices

(c) the foci of the hyperbola.

13. The equation of a hyperbola is $4x^2 - 9y^2 = 36$. Find the equation of the tangent to the hyperbola at the point $(3\sqrt{2}, 2)$.

14. The equation of a hyperbola is given by $x^2 - 4y^2 = 36$. Find the equation of the normal to the hyperbola at the point $(10, 4)$.

15. A hyperbola has foci $(\pm 17, 0)$ and eccentricity, $e = 2.125$. Find the equation of the hyperbola.

16. Classify the following equations into circle, parabola, ellipse and hyperbola

(a) $10y^2 - 2x^2 - 27y - 11x - 82 = 0$

(b) $4x^2 - 12y^2 - 20x + y - 21 = 0$

(c) $9x^2 - 20x + 24y - 21 = 0$

(d) $7x^2 + 7y^2 - 23x - 34y - 15 = 0$

(e) $5x^2 + 16y^2 + 54x - 32y - 25 = 0$

(f) $10x^2 + 10y^2 - x + 4y + 20 = 0$

(g) $5y^2 + 5x^2 + 12y + 5x + 35 = 0$

(h) $2y^2 + 16y - 20x + 24y - 21 = 0$

(i) $3y^2 + 7x^2 + 25y - 14x = 0$

(j) $20x^2 - 6y^2 - 74x + 125 = 0$

CHAPTER 8
LIMIT OF A FUNCTION

The concept of limits is very important in differential calculus. A function has a limiting value when its variable approaches a certain value. For example, if the limiting value of f(x) as x approaches 4 is 16, it is written as:

$$\lim_{x \to 4} f(x) = 16$$

In evaluating limit, some problems are simply solved by just putting in the values of the variables, while some problems are solved by applying certain rules.

Some important limits

1. $\lim_{x \to 0} \dfrac{\sin x}{x} = 1$ and $\lim_{x \to 0} \dfrac{\tan x}{x} = 1$ (Or $\lim_{x \to 0} \dfrac{\sin ax}{ax} = 1$ and $\lim_{x \to 0} \dfrac{\tan ax}{ax} = 1$)

2. $\lim_{x \to 0} \dfrac{1 - \cos x}{x} = 0$ and $\lim_{x \to 0} \dfrac{1 - \cos a\, x}{x} = 0$

3. $\lim_{x \to \infty} \left(1 + \dfrac{1}{x}\right)^x = e$

4. $\lim_{x \to 0} \dfrac{a^x - 1}{a^x} = 1$

5. $\lim_{x \to 0} \dfrac{e^x - 1}{x} = 1$

6. $\lim_{x \to 0} \dfrac{1 + x}{x} = 1$

7. $\lim_{x \to a} \dfrac{x^n - a^n}{x - a} = na^{n-1}$

8. $\lim_{x \to 0} \dfrac{a^x - 1}{x} = \ln a$

9. $\lim_{x \to \infty} \dfrac{\ln x}{x} = 0$

10. $\lim_{x \to \infty} x^{\frac{1}{x}} = 0$

11. $\lim_{x \to 0} (1 + x)^{\frac{1}{x}} = e$

12. $\lim_{x \to 0} (1 + \sin x)^{\frac{1}{x}} = e$

13. $\lim_{x \to a} c = c$

Examples

1. Evaluate $\lim_{x \to 0} 3x^3 - 5x^2 + 2x + 6$

Solution

$\lim_{x \to 0} 3x^3 - 5x^2 + 2x + 6$

Simply substitute zero for x in the given expression. This gives:

$= 3(0^3) - 5(0^2) + 2(0) + 6$

$= 0 - 0 + 0 + 6$

$= 6$

2. Evaluate $\lim_{x \to 0} \dfrac{x^2 + 3x - 11}{4x^2 - 5x - 5}$

Solution

$\lim_{x \to 0} \dfrac{x^2 + 3x - 11}{4x^2 - 5x - 5}$

Substituting 0 for x into the expression gives:

$= \dfrac{0^2 + 3(0) - 11}{4(0)^2 - 5(0) - 5}$

$= \dfrac{0 + 0 - 11}{0 - 0 - 5}$

$= \dfrac{-11}{-5}$

$= \dfrac{11}{5}$

$= 2\dfrac{1}{5}$

3. Evaluate $\lim_{x \to 4} \dfrac{x^2 - 2x - 8}{x - 4}$

Solution

A close look at the expression shows that if 4 is substituted for x, it will give $\dfrac{0}{0}$. This has no value as it is indeterminate. Therefore, in order to solve a limit such as this, we have to factorize the denominator and then simplify the expression. This is done as follows:

$\lim_{x \to 4} \dfrac{x^2 - 2x - 8}{x - 4} = \lim_{x \to 4} \dfrac{(x + 2)(x - 4)}{x - 4}$

Cancelling out $(x - 4)$ gives:

$\lim_{x \to 4} (x + 2)$

We now substitute 4 for x to obtain our final answer as follows:

$\lim_{x \to 4} (x + 2) = 4 + 2$

$= 6$

Therefore, $\lim_{x \to 4} \dfrac{x^2 + x - 8}{x - 4} = 6$

4. Evaluate $\lim\limits_{x \to 2} \dfrac{x^2 - 4}{x - 2}$

Solution

This is similar to example 3 above since substituting 2 for x in the expression will $\dfrac{0}{0}$ which has no value. Hence we factorize the numerator and simplify as follows:

$\lim\limits_{x \to 2} \dfrac{x^2 - 4}{x - 2} = \lim\limits_{x \to 2} \dfrac{(x-2)(x+2)}{x - 2}$ [Note that $x^2 - 4 = x^2 - 2^2$ and recall that $a^2 - b^2 = (a + b)(a - b)$]

Cancelling out $(x - 2)$ gives:

$\lim\limits_{x \to 2} (x + 2)$

We now substitute 2 for x to obtain our final answer as follows:

$\lim\limits_{x \to 2} (x + 2) = 2 + 2$

$= 4$

Therefore, $\lim\limits_{x \to 2} \dfrac{x^2 - 4}{x - 2} = 4$

5. Evaluate $\lim\limits_{x \to 0} (2x - 5)(x + 2)(3x - 1)$

Solution

$\lim\limits_{x \to 0} (2x - 5)(x + 2)(3x - 1)$

Substituting 0 for x in the expression gives:

$\lim\limits_{x \to 0} (2x - 5)(x + 2)(3x - 1) = [2(0) - 5][(0) + 2][3(0) - 1]$

$= (-5)(2)(-1)$

$= 10$

6. Evaluate $\lim\limits_{x \to \infty} \dfrac{x^3 + 2x^2 + 5x + 3}{x^3 + x^2 + 3x + 7}$

Solution

In this case of x approaching infinity, we evaluate it by first dividing each term in the numerator and denominator by the variable having the highest exponent (power). Hence we divide each term by x^3. This is done as follows:

$\lim\limits_{x \to \infty} \dfrac{x^3 + 2x^2 + 5x + 3}{x^3 + x^2 + 3x + 7} = \lim\limits_{x \to \infty} \dfrac{\frac{x^3}{x^3} + \frac{2x^2}{x^3} + \frac{5x}{x^3} + \frac{3}{x^3}}{\frac{x^3}{x^3} + \frac{x^2}{x^3} + \frac{3x}{x^3} + \frac{7}{x^3}}$

$= \lim\limits_{x \to \infty} \dfrac{1 + \frac{2}{x} + \frac{5}{x^2} + \frac{3}{x^3}}{1 + \frac{1}{x} + \frac{3}{x^2} + \frac{7}{x^3}}$

$= \dfrac{1 + \frac{2}{\infty} + \frac{5}{\infty^2} + \frac{3}{\infty^3}}{1 + \frac{1}{\infty} + \frac{3}{\infty^2} + \frac{7}{\infty^3}}$

$$= \frac{1 + 0 + 0 + 0}{1 + 0 + 0 + 0} \qquad \text{(Note that a number divided by } \infty \text{ gives 0)}$$

$$= \frac{1}{1}$$

$$= 1$$

7. Evaluate $\displaystyle\lim_{x \to \infty} \frac{2x^4 - 3x^2 + 8x - 1}{x^4 - 5x + 3}$

Solution

The variable in its highest exponent (power) is x^4. Hence we divide each term by x^4 as follows:

$$\lim_{x \to \infty} \frac{2x^4 - 3x^2 + 8x - 1}{x^4 - 5x + 3} = \lim_{x \to \infty} \frac{\dfrac{2x^4}{x^4} - \dfrac{3x^2}{x^4} + \dfrac{8x}{x^4} - \dfrac{1}{x^4}}{\dfrac{x^4}{x^4} - \dfrac{5x}{x^4} + \dfrac{3}{x^4}}$$

$$= \lim_{x \to \infty} \frac{2 - \dfrac{3}{x^2} + \dfrac{8}{x^3} - \dfrac{1}{x^4}}{1 - \dfrac{5}{x^3} + \dfrac{3}{x^4}}$$

$$= \frac{2 - \dfrac{3}{\infty} + \dfrac{8}{\infty} - \dfrac{1}{\infty}}{1 - \dfrac{5}{\infty} + \dfrac{3}{\infty}}$$

$$= \frac{2 - 0 + 0 - 0}{1 - 0 + 0} \qquad \text{(Note that a number divided by } \infty \text{ gives 0)}$$

$$= \frac{2}{1}$$

$$= 2$$

8. Evaluate $\displaystyle\lim_{x \to 4} \frac{2x^2 - 7x - 4}{x^2 - 3x - 4}$

Solution

If 4 is substituted for x in the function above, it will give $\dfrac{0}{0}$ which is an indeterminate value.

Hence we factorize the expression and simplify as follows:

$$\lim_{x \to 4} \frac{2x^2 - 7x - 4}{x^2 - 3x - 4} = \lim_{x \to 4} \frac{(2x+1)(x-4)}{(x+1)(x-4)}$$

Cancelling out $(x - 4)$ gives:

$$\lim_{x \to 4} \frac{2x + 1}{x + 1}$$

We now substitute 4 for x to obtain our final answer as follows:

$$\lim_{x \to 4} \frac{2x + 1}{x + 1} = \frac{2(4) + 1}{(4) + 1}$$

$$= \frac{8+1}{4+1}$$

$$= \frac{9}{5}$$

$$= 1\frac{4}{5}$$

Therefore, $\lim\limits_{x \to 4} \dfrac{2x^2 - 7x - 4}{x^2 - 3x - 4} = 1\dfrac{4}{5}$

9. Evaluate $\lim\limits_{x \to 3} \dfrac{x^2 - 9}{x^2 + 6x + 9}$

Solution

Substituting 3 for x in the function can give us the final answer as follows:

$$\lim\limits_{x \to 3} \frac{x^2 - 9}{x^2 + 6x + 9} = \frac{3^2 - 9}{3^2 + 6(3) + 9}$$

$$= \frac{9 - 9}{9 + 18 + 9}$$

$$= \frac{0}{36}$$

$$= 0$$

Hence, $\lim\limits_{x \to 3} \dfrac{x^2 - 9}{x^2 + 6x + 9} = 0$

Note that even if this expression had been factorized and simplified, the final answer would still give zero.

10. Determine the limiting value of $\dfrac{x + 2}{x^2 + 6x + 8}$ as x tends to -2 .

Solution

The question can also be written as $\lim\limits_{x \to -2} \dfrac{x + 2}{x^2 + 6x + 8}$

If -2 is substituted for x in the function above, it will give $\dfrac{0}{0}$. Hence we factorize the expression and simplify as follows:

$$\lim\limits_{x \to -2} \frac{x + 2}{x^2 + 6x + 8} = \lim\limits_{x \to -2} \frac{x + 2}{(x + 2)(x + 4)}$$

Cancelling out $(x + 2)$ gives:

$$\lim\limits_{x \to -2} \frac{1}{x + 4}$$

We now substitute -2 for x as follows:

$$\lim\limits_{x \to -2} \frac{1}{x + 4} = \frac{1}{-2 + 4}$$

$$= \frac{1}{2}$$

11. Find the limit of $\dfrac{x^3 - 125}{x - 5}$ as $x \to 5$

Solution

The question can also be written as $\lim\limits_{x \to 5} \dfrac{x^3 - 125}{x - 5}$

We factorize the expression and simplify as follows:

$\lim\limits_{x \to 5} \dfrac{x^3 - 125}{x - 5} = \lim\limits_{x \to 5} \dfrac{x^3 - 5^3}{x - 5}$

$\qquad\qquad = \lim\limits_{x \to 5} \dfrac{(x - 5)(x^2 + 5x + 5^2)}{x - 5}$ [Recall the identity $a^3 - b^3 = (a - b)(a^2 + ab + b^2)$]

Cancelling out $(x - 5)$ gives:

$\lim\limits_{x \to 5} (x^2 + 5x + 5^2)$

We now substitute 5 for x as follows:

$\lim\limits_{x \to 5} (x^2 + 5x + 5^2) = (5^2 + 5(5) + 25)$

$\qquad\qquad\qquad = 25 + 25 + 25$

$\qquad\qquad\qquad = 75$

12. Find the limit of $\dfrac{x^3 + 64}{x + 4}$ as $x \to -4$

Solution

The question can also be written as $\lim\limits_{x \to -4} \dfrac{x^3 + 4^3}{x + 4}$

We factorize the expression and simplify as follows:

$\lim\limits_{x \to -4} \dfrac{x^3 + 64}{x + 4} = \lim\limits_{x \to -4} \dfrac{x^3 + 4^3}{x + 4}$

$\qquad\qquad = \lim\limits_{x \to -4} \dfrac{(x + 4)(x^2 - 4x + 4^2)}{x + 4}$ [Recall the identity $a^3 + b^3 = (a + b)(a^2 - ab + b^2)$]

Cancelling out $(x + 4)$ gives:

$\lim\limits_{x \to -4} (x^2 - 4x + 4^2)$

We now substitute -4 for x as follows:

$\lim\limits_{x \to -4} (x^2 - 4x + 4^2) = (-4^2 - 4(-4) + 16)$

$\qquad\qquad\qquad = 16 + 16 + 16$

$\qquad\qquad\qquad = 48$

13. Evaluate $\lim\limits_{x \to 9} \dfrac{3 - \sqrt{x}}{9 - x}$

Solution

If 9 is substituted for x in the expression above, it will give $\dfrac{0}{0}$ which is not the desired answer.

Hence a way to go around this problem is to multiply the top and bottom by the conjugate of the surd at the top. The conjugate of $3 - \sqrt{x}$ is $3 + \sqrt{x}$ (only a difference in their middle signs).

143

Hence we multiply top and bottom by $3 + \sqrt{x}$ and simplify as follows:

$$\lim_{x \to 9} \frac{3 - \sqrt{x}}{9 - x} = \lim_{x \to 9} \frac{(3 - \sqrt{x})(3 + \sqrt{x})}{(9 - x)(3 + \sqrt{x})}$$

$$= \lim_{x \to 9} \frac{3^2 - (\sqrt{x})^2}{(9 - x)(3 + \sqrt{x})} \quad \text{(Note that the top was simplified by using } (a - b)(a + b) = a^2 - b^2)$$

$$= \lim_{x \to 9} \frac{9 - x}{(9 - x)(3 + \sqrt{x})}$$

$$= \lim_{x \to 9} \frac{1}{(3 + \sqrt{x})} \quad \text{(Since } 9 - x \text{ cancels out)}$$

We now substitute 9 for x as follows:

$$\lim_{x \to 9} \frac{1}{(3 + \sqrt{x})} = \frac{1}{(3 + \sqrt{9})}$$

$$= \frac{1}{3 + 3}$$

$$= \frac{1}{6}$$

14. Evaluate $\lim_{x \to 1} \dfrac{x^2 - 1}{x^5 - 1}$

Solution

If 2 is substituted for x in the expression above, it will give $\dfrac{0}{0}$ which is not a good answer. Hence a way to go around this problem is to divide the top and bottom by $x - 1$ (i.e. the difference between the variable and 1). This is done as follows:

$$\lim_{x \to 1} \frac{x^2 - 1}{x^5 - 1} = \lim_{x \to 1} \frac{\dfrac{x^2 - 1}{x - 1}}{\dfrac{x^5 - 1}{x - 1}} \quad \text{(Note that this has not changed the original fraction)}$$

Note that $\lim_{x \to a} \dfrac{x^n - a^n}{x - a} = na^{n-1}$. With this rule we now simplify the expression above as follows:

$$\lim_{x \to 1} \frac{\dfrac{x^2 - 1}{x - 1}}{\dfrac{x^5 - 1}{x - 1}} = \frac{2 \times 1^{(2-1)}}{5 \times 1^{(5-1)}}$$

$$= \frac{2 \times 1^1}{5 \times 1^5}$$

$$= \frac{2}{5}$$

Continuity of a function
A function is said to be continuous if the three conditions below are satisfied.

1. $f(a)$ exists

2. $\underset{x \to a}{lim}$ f(x) exists

3. $\underset{x \to a}{lim}$ f(x) = f(a)   (i.e. if the values of conditions 1 and 2 above are equal)

Examples

1. Determine if the function f(x) = $2x^2 - 8x + 5$ is continuous at the point $x = -1$

Solution

When $x = -1$, the f(−1) is obtained as follows:

f(x) = $2x^2 - 8x + 5$

f(−1) = $2(-1)^2 - 8(-1) + 5$

= 2 + 8 + 5

= 15

Since f(−1) = 15, it means that f(−1) exists.

The next step is to find the value of $\underset{x \to -1}{lim}$ f(x) as follows:

$\underset{x \to -1}{lim}$ $2x^2 - 8x + 5 = 2(-1)^2 - 8(-1) + 5$

= 2 + 8 + 5

= 15

Hence f(−1) = $\underset{x \to -1}{lim}$ f(x) = 15

Therefore, the function is continuous.

2. Determine if the function f(x) = $\dfrac{2x + 1}{x^2 - 3x + 7}$ is continuous at $x = 2$

Solution

When $x = 2$, the f(2) is obtained as follows:

f(x) = $\dfrac{2x + 1}{x^2 - 3x + 7}$

f(2) = $\dfrac{2(2) + 1}{2^2 - 3(2) + 7}$

= $\dfrac{4 + 1}{4 - 6 + 7}$

= $\dfrac{5}{5}$

= 1

Since f(2) = 1, it means that f(2) exists.

The next step is to find the value of $\underset{x \to 2}{lim}$ f(x) as follows:

$\underset{x \to 2}{lim}$ $\dfrac{2x + 1}{x^2 - 3x + 7}$ = $\dfrac{2(2) + 1}{2^2 - 3(2) + 7}$

= $\dfrac{4 + 1}{4 - 6 + 7}$

= $\dfrac{5}{5}$

$$= 1$$

Hence f(2) = $\lim_{x \to 2}$ f(x) = 1

Therefore, the function is continuous.

3. Determine if the function f(x) = $\dfrac{2x^2 - 8}{x + 4}$ is continuous at $x = -4$.

When $x = -4$, the f(-4) is obtained as follows:

$$f(x) = \frac{2x^2 - 8}{x + 4}$$

$$f(-4) = \frac{2(-4^2) - 8}{-4 + 4}$$

$$= \frac{2(16) - 8}{0}$$

$$= \frac{24}{0} \quad \text{(Undefined)}$$

Hence f(-4) is undefined, and does not exist. This shows that the function is not continuous. It is discontinuous as $x = -4$.

4. Determine if the function f(x) = $\dfrac{x^2 - 9}{x - 3}$ is continuous at $x = 3$.

When $x = 3$, the f(3) is obtained as follows:

$$f(x) = \frac{x^2 - 9}{x - 3}$$

$$f(3) = \frac{3^2 - 9}{3 - 3}$$

$$= \frac{0}{0} \quad \text{(This has no value)}$$

Hence f(3) does not exist.

Let us determine the value of $\lim_{x \to 3}$ f(x) as follows:

$$\lim_{x \to 3} \frac{x^2 - 9}{x - 3} = \lim_{x \to 3} \frac{(x - 3)(x + 3)}{x - 3}$$

$$= \lim_{x \to 3} (x + 3) \quad \text{(Since } x - 3 \text{ cancels out)}$$

$$= 3 + 3 \quad \text{(When 3 is substituted for } x)$$

$$= 6$$

Hence, f(3) has no value while $\lim_{x \to 3}$ f(x) = 6

This shows that f(3) $\neq$ $\lim_{x \to 3}$ f(x)

Therefore the function discontinuous.

5. Test for the continuity of the function: f(x) = $\begin{cases} -5x + 7 & \text{if } x > 1 \\ x^2 + 2 & \text{if } x \leq 1 \end{cases}$

Solution

$$f(x) = \begin{cases} -5x + 7 & \text{if } x > 1 \\ x^2 + 2 & \text{if } x \leq 1 \end{cases}$$

When $x = 1$, we use the expression $x^2 + 2$ to evaluate $f(1)$. Note that $x \leq 1$ means $x < 1$ or $x = 1$.

Hence, $f(x) = x^2 + 2$ (for $x = 1$)

$$f(1) = 1^2 + 2$$
$$= 3$$

Since $f(1) = 3$, it means that $f(1)$ exist.

Let us now determine the limiting value of $f(x)$.

If x approaches 1 from the left, which is $x \leq 1$ (left hand elbow) we write it as $\lim_{x \to 1-} f(x)$, and we pick the expression given by:

$$f(x) = x^2 + 2$$

Hence, $\lim_{x \to 1} f(x) = 1^2 + 3$

$$= 3$$

If x approaches 1 from the right, which is $x > 1$ (right hand elbow), we write it as $\lim_{x \to 1+} f(x)$, and we pick the expression given by:

$$f(x) = -5x + 7$$

Hence, $\lim_{x \to 1} f(x) = -5(1) + 7$

$$= 2$$

Hence, $\lim_{x \to 1-} f(x) = 3$ while $\lim_{x \to 1+} f(x) = 2$. They are not equal. This shows that $\lim_{x \to 1} f(x)$ does not exist. Therefore, the function is not continuous.

6. Determine the continuity of the function: $f(x) = \begin{cases} x^2 & \text{if } x \geq -2 \\ x+6 & \text{if } x < -2 \end{cases}$

Solution

$$f(x) = \begin{cases} x^2 & \text{if } x \geq -2 \\ x+6 & \text{if } x < -2 \end{cases}$$

A direct way of determining the continuity of functions such as these is to find the limits of the two expressions in the function as x tends to the values given in the question.

Therefore, when x tends to -2 in the function $f(x) = x^2$, we have:

$$f(x) = x^2$$

Hence, $\lim_{x \to -2} f(x) = (-2)^2$

$$= 4$$

Also, when x tends to -2 in the function $f(x) = x + 6$, we have:

$$f(x) = x + 6$$

Hence, $\lim_{x \to -2} f(x) = -2 + 6$

$$= 4$$

Hence, $\lim_{x \to -2} x^2 = \lim_{x \to -2} x + 6 = 4$. Their limits give the same value. This shows that $\lim_{x \to -2} f(x)$ exist. Therefore, the function is continuous.

147

7. Determine if the function below is continuous or not.

$$f(x) = \begin{cases} \dfrac{x^2 - 16}{x + 4} & \text{if } x \geq -4 \\ \dfrac{x + 4}{2x - 3} & \text{if } x < -4 \end{cases}$$

Solution

As x tends to -4 in the function $f(x) = \dfrac{x^2 - 16}{x + 4}$ we simplify as follows:

$$f(x) = \frac{(x + 4)(x - 4)}{x + 4}$$

Hence, $\lim\limits_{x \to -4} f(x) = (x - 4)$ (Note that $(x + 4)$ cancels out)

$$= -4 - 4$$
$$= -8$$

Also, when x tends to -4 in the function $f(x) = \dfrac{x + 4}{2x - 3}$, we have:

$$\lim\limits_{x \to -4} f(x) = \frac{-4 + 4}{2(-4) - 3}$$
$$= \frac{-4 + 4}{-8 - 3}$$
$$= \frac{0}{-11}$$
$$= 0$$

Their limits give different values. This shows that $\lim\limits_{x \to -4} f(x)$ does not exist. Therefore, the function is not continuous.

8. If $f(x) = \begin{cases} \dfrac{x^2 - 4}{x - 2} \\ 4 \ \text{if } x = 2 \end{cases}$ test for the continuity of the function at $x = 2$.

Solution

$$f(x) = \begin{cases} \dfrac{x^2 - 4}{x - 2} \\ 4 \ \text{if } x = 2 \end{cases}$$

Let us find the limits of the two expressions in the function as x tends 2.

From the first expression, as x tends to 2 we have:

$$f(x) = \frac{x^2 - 4}{x - 2}$$

Hence, $\lim\limits_{x \to 2} f(x) = \dfrac{(x - 2)(x + 2)}{x - 2}$ (When the numerator is factorized)

$$\lim\limits_{x \to 2} f(x) = (x + 2) \quad (x - 2 \text{ has cancelled out})$$
$$\lim\limits_{x \to 2} f(x) = 2 + 2$$
$$= 4$$

Also, when x tends to 2 in the function $f(x) = 4$, we have:

$$f(x) = 4$$

148

Hence, $\underset{x \to 2}{lim} \, 4 = 4$ (Since $\underset{x \to a}{lim} \, c = c$)

The question also tells us that when $x = 2$, $f(x) = 4$

Hence, $\underset{x \to 2}{lim} \, 4 = \underset{x \to 2}{lim} \, \dfrac{x^2 - 4}{x - 2} = 4$. Their limits give the same value of 4. This shows that f(x) exist,

and that $f(2) = \underset{x \to 2}{lim} \, f(x)$. Therefore, the function is continuous.

Examples on Limits of Trigonometric Functions

1. Evaluate $\underset{x \to 0}{lim} \, \dfrac{\sin 2x}{5x}$

Solution

$\underset{x \to 0}{lim} \, \dfrac{\sin 2x}{5x}$

Substituting zero directly into the expression above will give $\dfrac{0}{0}$. Hence we have to apply the

appropriate rule of limit. Let us first make an adjustment to the expression as follows.

Multiply the expression by $\dfrac{2x}{5x}$ and change the denominator to $2x$ just like the numerator, as

follows:

$\underset{x \to 0}{lim} \, \dfrac{\sin 2x}{5x} = \underset{x \to 0}{lim} \, \left(\dfrac{\sin 2x}{2x} \times \dfrac{2x}{5x} \right)$

Note that when $2x$ cancels out, the original expression remains the same.

Recall that $\underset{x \to 0}{lim} \, \dfrac{\sin ax}{ax} = 1$

Hence, $\underset{x \to 0}{lim} \, \dfrac{\sin 2x}{2x} = 1$

Substituting 1 for $\underset{x \to 0}{lim} \, \dfrac{\sin 2x}{2x}$ in the simplification given above to gives:

$$\underset{x \to 0}{lim} \, \left(\dfrac{\sin 2x}{2x} \times \dfrac{2x}{5x} \right) = \underset{x \to 0}{lim} \, \dfrac{\sin 2x}{2x} \times \underset{x \to 0}{lim} \, \dfrac{2x}{5x}$$

$$= 1 \times \underset{x \to 0}{lim} \, \dfrac{2}{5} \quad (x \text{ cancels out})$$

$$= 1 \times \dfrac{2}{5} \quad (\text{Since } \underset{x \to a}{lim} \, c = c)$$

$$= \dfrac{2}{5}$$

2. 1. Evaluate $\underset{x \to 0}{lim} \, \dfrac{\sin 5x}{7x}$

Solution

$\underset{x \to 0}{lim} \, \dfrac{\sin 5x}{7x}$

Multiply the expression by $\dfrac{5x}{7x}$ and change the denominator to $5x$ just like the numerator. This

gives:

$$\lim_{x \to 0} \frac{\sin 5x}{7x} = \lim_{x \to 0} \left(\frac{\sin 5x}{5x} \times \frac{5x}{7x} \right)$$

Hence, $\lim_{x \to 0} \frac{\sin 5x}{5x} = 1$

Substituting 1 for $\lim_{x \to 0} \frac{\sin 5x}{5x}$ in the simplification given above to gives:

$$\lim_{x \to 0} \left(\frac{\sin 5x}{5x} \times \frac{5x}{7x} \right) = \lim_{x \to 0} \frac{\sin 5x}{5x} \times \lim_{x \to 0} \frac{5x}{7x}$$

$$= 1 \times \lim_{x \to 0} \frac{5}{7} \quad (x \text{ cancels out})$$

$$= 1 \times \frac{5}{7} \quad (\text{Since } \lim_{x \to a} c = c)$$

$$= \frac{5}{7}$$

The two examples above show that $\lim_{x \to 0} \frac{\sin a\, x}{x} = a$ or $\lim_{x \to 0} \frac{\sin a\, x}{bx} = \frac{a}{b}$. This rule also applies to tangent as $\lim_{x \to 0} \frac{\tan a\, x}{x} = a$ or $\lim_{x \to 0} \frac{\tan a\, x}{bx} = \frac{a}{b}$.

3. Find the value of $\lim_{x \to \frac{\pi}{2}} \frac{\cos 2x}{\sin 3x}$

Solution

$$\lim_{x \to \frac{\pi}{2}} \frac{\cos 2x}{\sin 3x}$$

We can simply substitute in the value of x and obtain our answer.

$$\lim_{x \to \frac{\pi}{2}} \frac{\cos 2x}{\sin 3x} = \frac{\cos 2\left(\frac{\pi}{2}\right)}{\sin 3\left(\frac{\pi}{2}\right)}$$

$$= \frac{\cos \pi}{\sin \frac{3\pi}{2}}$$

$$= \frac{\cos 180}{\sin 270} \quad (\text{Note that } \pi \text{ radians} = 180^\circ)$$

$$= \frac{-1}{-1}$$

$$= 1$$

This problem can also be solved by using the angles directly in radians, but I prefer to work in degrees. Note that cos π in radians = –1, and $\sin \frac{3\pi}{2}$ in radians = –1 as obtained above.

4. Evaluate $\lim_{x \to 0} \frac{\cos x}{\sin x - 3}$

Solution

$$\lim_{x \to 0} \frac{\cos x}{\sin x - 3} = \frac{\cos 0}{\sin 0 - 3}$$

$$= \frac{1}{0-3}$$

$$= -\frac{1}{3}$$

5. Evaluate $\lim_{x \to 0} \frac{\sin 3x}{\sin 2x}$

Solution

Multiply by $\frac{x}{\sin 2x}$ and change the denominator to x as follows.

$$\lim_{x \to 0} \frac{\sin 3x}{\sin 2x} = \lim_{x \to 0} \left(\frac{\sin 3x}{x} \ \text{X} \ \frac{x}{\sin 2x} \right) \qquad \text{(When } x \text{ cancels out it gives the original expression)}$$

$$= \lim_{x \to 0} \left(\frac{\sin 3x}{x} \ \text{X} \ \left(\frac{\sin 2x}{x} \right)^{-1} \right) \qquad \text{(Note that the inverse of } \frac{x}{\sin 2x} \text{ has been taken)}$$

$$= \lim_{x \to 0} \frac{\sin 3x}{x} \ \text{X} \ \left(\lim_{x \to 0} \frac{\sin 2x}{x} \right)^{-1}$$

$$= 3 \times 2^{-1} \qquad \text{(Note that } \lim_{x \to 0} \frac{\sin 3x}{x} = 3 \ \text{ and } \ \lim_{x \to 0} \frac{\sin 2x}{x} = 2 \text{)}$$

$$= 3 \times \frac{1}{2}$$

$$= \frac{3}{2}$$

This example shows that: $\lim_{x \to 0} \frac{\sin a\, x}{\sin b\, x} = \frac{a}{b}$

6. Evaluate $\lim_{x \to 0} \frac{\sin 2x}{\sin 5x}$

Solution

$$\lim_{x \to 0} \frac{\sin 2x}{\sin 5x}$$

From the rule established in example 5 above, we can see that the solution to this problem is $\frac{2}{5}$.

Hence, $\lim_{x \to 0} \frac{\sin 2x}{\sin 5x} = \frac{2}{5}$ \quad (Since $\lim_{x \to 0} \frac{\sin a\, x}{\sin b\, x} = \frac{a}{b}$)

Exercise 8

1. Evaluate $\lim_{x \to 0} 2x^3 - 4x^2 + x + 9$

2. Evaluate $\lim_{x \to 0} \frac{5x^2 + x - 8}{x^2 - 2x + 5}$

3. Evaluate $\lim_{x \to 5} \frac{x^2 - x - 20}{x - 5}$

4. Evaluate $\lim_{x \to 7} \frac{x^2 - 49}{x - 7}$

5. Evaluate $\lim\limits_{x \to 0} (5x - 1)(3x + 7)(3x + 2)$

6. Evaluate $\lim\limits_{x \to \infty} \dfrac{2x^3 + x^2 - 9x - 5}{5x^3 + 2x^2 - 7x + 2}$

7. Evaluate $\lim\limits_{x \to \infty} \dfrac{x^4 - 5x^2 - 2x - 6}{3x^4 + 2x + 3}$

8. Evaluate $\lim\limits_{x \to 4} \dfrac{x^2 - 25}{x^2 + 2x - 15}$

9. Evaluate $\lim\limits_{x \to 8} \dfrac{2x^2 - 17x + 8}{8 - x}$

10. Evaluate $\lim\limits_{m \to 0} \dfrac{(5+m)^2 - 25}{m}$

11. Determine the limiting value of $\dfrac{\sqrt{x} - 2}{x - 4}$ as x tends to 4 .

12. Find the limit of $\dfrac{x^3 - 27}{x - 3}$ as $x \to 3$

13. Find the limit of $\dfrac{x^3 + 8}{x + 2}$ as $x \to -2$

14. Evaluate $\lim\limits_{x \to 25} \dfrac{5 - \sqrt{x}}{25 - x}$

15. Evaluate $\lim\limits_{x \to 1} \dfrac{x^3 - 1}{x^7 - 1}$

16. Determine if the function $f(x) = x^2 + 3x - 1$ is continuous at the point $x = 2$

17. Determine if the function $f(x) = \dfrac{5x - 2}{3x^2 - x - 2}$ is continuous at $x = -1$

18. Determine if the function $f(x) = \dfrac{2x^2 - 18}{x - 3}$ is continuous at $x = 3$.

19. Determine if the function $f(x) = \dfrac{x^3 + 64}{x + 4}$ is continuous at $x = -4$.

20. Test for the continuity of the function: $f(x) = \begin{cases} -x + 5 & \text{if } x > 2 \\ x^3 + 10 & \text{if } x \le 2 \end{cases}$

21. Determine the continuity of the function: $f(x) = \begin{cases} 3x^2 & \text{if } x \ge -3 \\ x + 30 & \text{if } x < -3 \end{cases}$

22. Determine if the function below is continuous or not.

$$f(x) = \begin{cases} \dfrac{x^2 - 25}{x + 5} & \text{if } x \ge -5 \\ \dfrac{x + 5}{2x + 10} & \text{if } x < -5 \end{cases}$$

23. If $f(x) = \begin{cases} \dfrac{x^3 - 27}{x - 3} \\ 1 \quad \text{if } x = 3 \end{cases}$ test for the continuity of the function at $x = 3$.

24. If $f(x) = \begin{cases} 2 - 3x & x < 1 \\ x^3 + 4 & x \ge 1 \end{cases}$ evaluate the following limits, if they exist.

(a) $\lim\limits_{x \to -2} f(x)$

(b) $\lim\limits_{x \to 1} f(x)$

25. Evaluate $\lim\limits_{x \to 2} (5 + |x - 2|)$ if it exists.

26. Evaluate $\lim\limits_{x \to 0} \dfrac{\sin 7x}{2x}$

27. Evaluate $\lim\limits_{x \to 0} \dfrac{\sin x}{3x}$

28. Find the value of $\lim\limits_{x \to \pi} \dfrac{\cos 2x}{\sin \frac{x}{2}}$

29. Evaluate $\lim\limits_{x \to 0} \dfrac{\cos 5x}{\sin 3x - 2}$

30. Evaluate $\lim\limits_{x \to 0} \dfrac{\sin 9x}{\sin 4x}$

CHAPTER 9

DIFFERENTIATION FROM FIRST PRINCIPLE

If y = f(x), then the gradient function of y or f(x) is given by:

$$g(x) = \frac{f(x + \Delta x) - f(x)}{\Delta x}$$

where Δx is the increment in x.

Applying limits has shown that as Δx tends to zero,

$$f'(x) = \lim_{\Delta x \to 0} \frac{f(x + \Delta x) - f(x)}{\Delta x}$$

$$\frac{dy}{dx} = \lim_{\Delta x \to 0} \frac{f(x + \Delta x) - f(x)}{\Delta x}$$

$\dfrac{dy}{dx}$ reads dee y dee x.

Note that f'(x) and $\dfrac{dy}{dx}$ can be interpreted as instantaneous rate of change of y with respect to x.

The technique of finding the derivative of a function by considering the limiting value is called differentiation from first principle.

Examples

1. Find the derivative of x from first principle.

Solution

Let, y = x

Increasing x by Δx will give an increase in y by Δy. This gives:

$$y + \Delta y = x + \Delta x$$
$$\Delta y = x + \Delta x - y$$
$$\Delta y = x + \Delta x - x \quad \text{(substitute } x \text{ for y since y} = x \text{ as given above from the question)}$$
$$\Delta y = \Delta x \quad (x \text{ cancels out)}$$

Dividing both sides by Δx in order to get the derivative gives:

$$\frac{\Delta y}{\Delta x} = \frac{\Delta x}{\Delta x}$$
$$\frac{\Delta y}{\Delta x} = 1$$

Taking limits as Δx tends to zero gives:

$$\lim_{\Delta x \to 0} \frac{\Delta y}{\Delta x} = 1 \quad [\text{Recall that } \lim_{x \to 0} c = c, \text{ hence } \lim_{\Delta x \to 0} f(\Delta x) = 1 \text{ gives 1, since } \frac{\Delta y}{\Delta x} \text{ is regarded as f}(\Delta x)]$$

We now write $\lim_{\Delta x \to 0} \dfrac{\Delta y}{\Delta x}$ as $\dfrac{dy}{dx}$. This gives:

$$\frac{dy}{dx} = 1$$

2. Find from first principle, the derivative of $2x^3$.

Solution

Let, $y = 2x^3$

Increasing x by Δx will give an increase in y by Δy. This gives:

$\quad y + \Delta y = 2(x + \Delta x)^3$

$\quad y + \Delta y = 2[x^3 + 3x^2\Delta x + 3x(\Delta x)^2 + (\Delta x)^3]$

$\quad\quad = 2x^3 + 6x^2\Delta x + 6x(\Delta x)^2 + 2(\Delta x)^3$

$\quad\quad \Delta y = 2x^3 + 6x^2\Delta x + 6x(\Delta x)^2 + 2(\Delta x)^3 - y$

$\quad\quad\quad = 2x^3 + 6x^2\Delta x + 6x(\Delta x)^2 + 2(\Delta x)^3 - 2x^3$ $\quad$ (Since $y = 2x^3$)

$\quad\quad \Delta y = 6x^2\Delta x + 6x(\Delta x)^2 + 2(\Delta x)^3$ $\quad$ (Since $2x^3$ cancels out)

Dividing both sides by Δx in order to get the derivative gives:

$$\frac{\Delta y}{\Delta x} = \frac{6x^2\Delta x}{\Delta x} + \frac{6x(\Delta x)^2}{\Delta x} + \frac{2(\Delta x)^3}{\Delta x}$$

$$\frac{\Delta y}{\Delta x} = 6x^2 + 6x\Delta x + 2(\Delta x)^2$$

Taking limits as Δx tends to zero gives:

$$\lim_{\Delta x \to 0} \frac{\Delta y}{\Delta x} = 6x^2 + 6x(0) + 2(0)^2 \quad \text{(Note that zero replaces } \Delta x)$$

$$\lim_{\Delta x \to 0} \frac{\Delta y}{\Delta x} = 6x^2$$

We now write $\lim_{\Delta x \to 0} \frac{\Delta y}{\Delta x}$ as $\frac{dy}{dx}$. This gives:

$$\frac{dy}{dx} = 6x^2$$

3. Find the derivative of $f(x) = \dfrac{1}{x^2}$ from first principle.

Solution

$\quad f(x) = \Delta x$

Increasing x by Δx gives:

$$f(x + \Delta x) = \frac{1}{(x + \Delta x)^2}$$

$$f(x + \Delta x) = \frac{1}{x^2 + 2x\Delta x + (\Delta x)^2}$$

Subtract $f(x)$ from both sides of the equation

$$f(x + \Delta x) - f(x) = \frac{1}{x^2 + 2x\Delta x + (\Delta x)^2} - f(x)$$

$$f(x + \Delta x) - f(x) = \frac{1}{x^2 + 2x\Delta x + (\Delta x)^2} - \frac{1}{x^2} \quad \text{(Note that } f(x) = \frac{1}{x^2})$$

$$= \frac{x^2 - (x^2 + 2x\Delta x + (\Delta x)^2)}{x^2(x^2 + 2x\Delta x + (\Delta x)^2)}$$

$$= \frac{x^2 - x^2 - 2x\Delta x - (\Delta x)^2}{x^2(x^2 + 2x\Delta x + (\Delta x)^2)}$$

$$= \frac{-2x\Delta x - (\Delta x)^2}{x^2(x^2 + 2x\Delta x + (\Delta x)^2)}$$

$$f(x + \Delta x) - f(x) = \frac{\Delta x(-2x - \Delta x)}{x^2(x^2 + 2x\Delta x + (\Delta x)^2)}$$

Dividing both sides by Δx gives:

$$\frac{f(x + \Delta x) - f(x)}{\Delta x} = \frac{-2x - \Delta x}{x^2(x^2 + 2x\Delta x + (\Delta x)^2)} \qquad (\Delta x \text{ has cancelled out from the right hand side})$$

Taking limits as Δx tends to zero gives:

$$\lim_{\Delta x \to 0} \frac{f(x + \Delta x) - f(x)}{\Delta x} = \frac{-2x - 0}{x^2(x^2 + 2x(0) + (0)^2)}$$

Replacing $\displaystyle\lim_{\Delta x \to 0} \frac{f(x + \Delta x) - f(x)}{\Delta x}$ with f'(x) gives the derivative of f(x) as:

$$f'(x) = \frac{-2x}{x^2(x^2)}$$

$$f'(x) = \frac{-2}{x^3}$$

4. If $f(x) = 3x^2$

(a) write down and simplify the expression $\dfrac{f(x + h) - f(x)}{h}$ $\quad (h \neq 0)$

(b) find $\displaystyle\lim_{\Delta x \to 0} \frac{f(x + h) - f(x)}{h}$

Solution

(a) Let, $f(x) = 3x^2$

Increasing x by h gives:

$$f(x + h) = 3(x + h)^2$$
$$= 3[x^2 + 2xh + h^2]$$
$$= 3x^2 + 6xh + 3h^2$$

Subtracting f(x) from both sides gives:

$$f(x + h) - f(x) = 3x^2 + 6xh + 3h^2 - 3x^2 \qquad \text{(Note that } f(x) = 3x^2 \text{ from the question)}$$
$$= 6xh + 3h^2$$

Dividing both side by h gives:

$$\frac{f(x + h) - f(x)}{h} = \frac{6xh}{h} + \frac{3h^2}{h}$$

$$\frac{f(x + h) - f(x)}{h} = 6x + 3h$$

(b) $\quad \dfrac{f(x + h) - f(x)}{h} = 6x + 3h$

Taking limits as h tends to zero gives:

$$\lim_{h \to 0} \frac{f(x + h) - f(x)}{h} = 6x + 3(0)$$

$$\lim_{h \to 0} \frac{f(x + h) - f(x)}{h} = 6x$$

5. If $f(x) = 4x^3 - 5$,

(a) evaluate $\dfrac{f(x+h) - f(x)}{h}$, where $h \neq 0$

(b) From your result in (5a) above , find the derivatives of $f(x)$ with respect to x.

Solution

(a) Let, $f(x) = 4x^3 - 5$

Increasing x by h gives:

$$f(x + h) = 4(x + h)^3 - 5$$
$$= 4[x^3 + 3x^2h + 3xh^2 + h^3] - 5$$
$$= 4x^3 + 12x^2h + 12xh^2 + 4h^3 - 5$$

Subtracting $f(x)$ from both sides gives:

$$f(x + h) - f(x) = 4x^3 + 12x^2h + 12xh^2 + 4h^3 - 5 - (4x^3 - 5) \quad (f(x) = 4x^3 - 5 \text{ from the question})$$
$$= 4x^3 + 12x^2h + 12xh^2 + 4h^3 - 5 - 4x^3 + 5$$
$$= 12x^2h + 12xh^2 + 4h^3$$

Dividing both side by h gives:

$$\frac{f(x+h) - f(x)}{h} = 12x^2 + 12xh + 4h^2$$

(b) $\dfrac{f(x+h) - f(x)}{h} = 12x^2 + 12xh + 4h^2$

Taking limits as h tends to zero gives:

$$\lim_{h \to 0} \frac{f(x+h) - f(x)}{h} = 12x^2 + 12x(0)\ 4(0)^2$$

$$\lim_{h \to 0} \frac{f(x+h) - f(x)}{h} = 12x^2$$

6. If $f(x) = \dfrac{x^2 + 1}{2x}$

(a) write down and simplify the expression $\dfrac{f(x + \Delta x) - f(\Delta x)}{\Delta x}$, where $\Delta x \neq 0$

(b) Find the derivatives of $y = \dfrac{x^2 + 1}{2x}$

Solution

$$f(x) = \frac{x^2 + 1}{2x}$$

$$f(x + \Delta x) = \frac{(x + \Delta x)^2 + 1}{2(x + \Delta x)}$$

$$f(x + \Delta x) = \frac{x^2 + 2x\Delta x + (\Delta x)^2 + 1}{2x + 2\Delta x}$$

Subtract $f(x)$ from both sides of the equation

$$f(x + \Delta x) - f(x) = \frac{x^2 + 2x\Delta x + (\Delta x)^2 + 1}{2x + 2\Delta x} - \frac{x^2 + 1}{2x}$$

157

$$f(x + \Delta x) - f(x) = \frac{2x[x^2 + 2x\Delta x + (\Delta x)^2 + 1] - (x^2 + 1)(2x + 2\Delta x)}{2x(2x + 2\Delta x)}$$

$$= \frac{2x^3 + 4x^2\Delta x + 2x(\Delta x)^2 + 2x - (2x^3 + 2x^2\Delta x + 2x + 2\Delta x)}{2x(2x + 2\Delta x)}$$

$$= \frac{2x^3 + 4x^2\Delta x + 2x(\Delta x)^2 + 2x - 2x^3 - 2x^2\Delta x - 2x - 2\Delta x}{2x(2x + 2\Delta x)}$$

$$= \frac{2x^2\Delta x + 2x(\Delta x)^2 - 2\Delta x}{2x(2x + 2\Delta x)}$$

$$f(x + \Delta x) - f(x) = \frac{\Delta x(2x^2 + 2x\Delta x - 2)}{2x(2x + 2\Delta x)} \qquad \text{(After factorizing the numerator)}$$

Dividing both sides by Δx gives:

$$\frac{f(x + \Delta x) - f(x)}{\Delta x} = \frac{2x^2 + 2x\Delta x - 2}{2x(2x + 2\Delta x)}$$

(b) Taking limits as Δx tends to zero gives:

$$\lim_{\Delta x \to 0} \frac{f(x + \Delta x) - f(x)}{\Delta x} = \frac{2x^2 + 2x(0) - 2}{2x(2x + 2(0))}$$

$$= \frac{2x^2 - 2}{2x(2x)}$$

$$= \frac{2(x^2 - 1)}{4x^2}$$

Replacing $\lim_{\Delta x \to 0} \frac{f(x + \Delta x) - f(x)}{\Delta x}$ with f'(x) gives the derivative of f(x) as:

$$f'(x) = \frac{x^2 - 1}{2x^2}$$

Separating into fractions gives:

$$f'(x) = \frac{x^2}{2x^2} - \frac{1}{2x^2}$$

$$f'(x) = \frac{1}{2} - \frac{1}{2x^2}$$

7. Find from first principle the derivatives with respect to x of $y = 5x^3 - x + 7$

Solution

$$y = 5x^3 - x + 7$$

$$y + \Delta y = 5(x + \Delta x)^3 - (x + \Delta x) + 7$$

$$y + \Delta y = 5[x^3 + 3x^2\Delta x + 3x(\Delta x)^2 + (\Delta x)^3] - x - \Delta x + 7$$

$$= 5x^3 + 15x^2\Delta x + 15x(\Delta x)^2 + 5(\Delta x)^3 - x - \Delta x + 7$$

$$\Delta y = 5x^3 + 15x^2\Delta x + 15x(\Delta x)^2 + 5(\Delta x)^3 - x - \Delta x + 7 - y$$

$$= 5x^3 + 15x^2\Delta x + 15x(\Delta x)^2 + 5(\Delta x)^3 - x - \Delta x + 7 - (5x^3 - x + 7) \quad \text{(Since } y = 5x^3 - x + 7\text{)}$$

$$\Delta y = 5x^3 + 15x^2\Delta x + 15x(\Delta x)^2 + 5(\Delta x)^3 - x - \Delta x + 7 - 5x^3 + x - 7$$

$$= 15x^2\Delta x + 15x(\Delta x)^2 + 5(\Delta x)^3 - \Delta x$$

Dividing both sides by Δx gives:

$$\frac{\Delta y}{\Delta x} = \frac{15x^2\Delta x}{\Delta x} + \frac{15x(\Delta x)^2}{\Delta x} + \frac{5(\Delta x)^3}{\Delta x} - \frac{\Delta x}{\Delta x}$$

$$\frac{\Delta y}{\Delta x} = 15x^2 + 15x\Delta x + 5(\Delta x)^2 - 1$$

Taking limits as Δx tends to zero gives:

$$\lim_{\Delta x \to 0} \frac{\Delta y}{\Delta x} = 15x^2 + 15x(0) + 5(0)^2 - 1$$

$$\lim_{\Delta x \to 0} \frac{\Delta y}{\Delta x} = 15x^2 - 1$$

We now replace $\lim_{\Delta x \to 0} \frac{\Delta y}{\Delta x}$ with $\frac{dy}{dx}$. This gives:

$$\frac{dy}{dx} = 15x^2 - 1$$

8. Differentiate $3x - \frac{1}{2x^2}$ from first principle.

Solution

$$y = 3x - \frac{1}{x^2}$$

$$y + \Delta y = 3(x + \Delta x) - \frac{1}{2(x + \Delta x)^2}$$

$$= 3x + 3\Delta x - \frac{1}{2[x^2 + 2x\Delta x + (\Delta x)^2]}$$

Subtract y from both sides of the equation

$$\Delta y = 3x + 3\Delta x - \frac{1}{2x^2 + 4x\Delta x + 2(\Delta x)^2} - y$$

$$= 3x + 3\Delta x - \frac{1}{2x^2 + 4x\Delta x + 2(\Delta x)^2} - (3x - \frac{1}{2x^2})$$

$$= 3x + 3\Delta x - \frac{1}{2x^2 + 4x\Delta x + 2(\Delta x)^2} - 3x + \frac{1}{2x^2}$$

$$= 3\Delta x - \frac{1}{2x^2 + 4x\Delta x + 2(\Delta x)^2} + \frac{1}{2x^2}$$

Combining them into one fraction gives:

$$\Delta y = \frac{3\Delta x[2x^2(2x^2 + 4x\Delta x + 2(\Delta x)^2] - 2x^2 + 2x^2 + 4x\Delta x + 2(\Delta x)^2}{2x^2[2x^2 + 4x\Delta x + 2(\Delta x)^2]}$$

$$= \frac{3\Delta x[4x^4 + 8x^3\Delta x + 4x^2(\Delta x)^2] + 4x\Delta x + 2(\Delta x)^2}{2x^2[2x^2 + 4x\Delta x + 2(\Delta x)^2]} \quad (-2x^2 + 2x^2 \text{ has cancelled out})$$

$$= \frac{12x^4\Delta x + 24x^3(\Delta x)^2 + 12x^2(\Delta x)^3 + 4x\Delta x + 2(\Delta x)^2}{2x^2[2x^2 + 4x\Delta x + 2(\Delta x)^2]}$$

Factorizing the numerator gives:

$$= \frac{\Delta x[12x^4 + 24x^3\Delta x + 12x^2(\Delta x)^2 + 4x + 2\Delta x]}{2x^2[2x^2 + 4x\Delta x + 2(\Delta x)^2]}$$

Dividing both sides by Δx gives:

$$\frac{\Delta y}{\Delta x} = \frac{12x^4 + 24x^3\Delta x + 12x^2(\Delta x)^2 + 4x + 2\Delta x}{2x^2[2x^2 + 4x\Delta x + 2(\Delta x)^2]} \qquad (\Delta x \text{ that is outside the bracket cancels out})$$

Taking limits as Δx tends to zero gives:

$$\lim_{\Delta x \to 0}\frac{\Delta y}{\Delta x} = \frac{12x^4 + 24x^3(0) + 12x^2(0)^2 + 4x + 2(0)}{2x^2[2x^2 + 4x(0) + 2(0)^2]}$$

$$= \frac{12x^4 + 4x}{2x^2[2x^2]}$$

$$\lim_{\Delta x \to 0}\frac{\Delta y}{\Delta x} = \frac{12x^4 + 4x}{4x^4}$$

$$= \frac{12x^4}{4x^4} + \frac{4x}{4x^4}$$

$$= 3 + \frac{1}{x^3}$$

Replacing $\lim\limits_{\Delta x \to 0}\dfrac{\Delta y}{\Delta x}$ with $\dfrac{dy}{dx}$ gives the derivative as:

$$\frac{dy}{dx} = 3 + \frac{1}{x^3}$$

Exercise 9

1. Find the derivative of $2x$ from first principle.

2. Find from first principle, the derivative of x^2.

3. Find the derivative of $f(x) = \dfrac{1}{x^3}$ from first principle.

4. If $f(x) = 5x^2$

(a) write down and simplify the expression $\dfrac{f(x+h) - f(x)}{h}$ $\quad$ (h $\neq$ 0)

(b) find $\lim\limits_{h \to 0}\dfrac{f(x+h) - f(x)}{h}$

5. If $f(x) = 9x^3$,

(a) evaluate $\dfrac{f(x+h) - f(x)}{h}$, where h $\neq$ 0

(b) From your result above , find the derivatives of $f(x)$ with respect to x.

6. If $f(x) = \dfrac{x^3 - 2}{x}$

(a) write down and simplify the expression $\dfrac{f(x + \Delta x) - f(\Delta x)}{\Delta x}$, where $\Delta x \neq 0$

(b) Find the derivatives of $y = \dfrac{x^3 - 2}{x}$

7. Find from first principle the derivatives with respect to x of $y = 3x^2 - 10x$

8. Differentiate $x + \dfrac{3}{x}$ from first principle.

9. From first principle, find the derivatives with respect to x of $y = 5x - 3x^2$

10. Differentiate $2x + \dfrac{x}{5}$ from first principle.

CHAPTER 10

GENERAL RULE OF DIFFERENTIATION AND COMPOSITE FUNCTIONS

The general rule for the derivative/differentiation of a function is as given below.

If $y = x^n$

then, $\dfrac{dy}{dx} = nx^{n-1}$

The rule for differentiating a composite function (or function of a function) is given as follows:

If $y = (2x + 5)^4$

then we write, $u = 2x + 5$

and express y as:

$y = u^4$

Therefore, $\dfrac{dy}{dx} = \dfrac{dy}{du} \times \dfrac{du}{dx}$

This rule is called the chain rule.

Examples

1. Find the derivatives of the following:

(a) $y = 2x^7$

(b) $y = \dfrac{3}{4}x^8$

(c) $y = 3\sqrt{x}$

(d) $y = \dfrac{10}{\sqrt[5]{x^3}}$

(e) $y = \dfrac{1}{x^{\frac{1}{4}}}$

Solution

(a) $y = 2x^7$

Solutions

(a) $y = 2x^7$

$\dfrac{dy}{dx} = 7 \times 2x^{7-1}$

This is done by multiplying the exponent (power) by the term and subtracting 1 from the exponent (power). Hence, the answer is:

$\dfrac{dy}{dx} = 14x^6$

(b) $y = \dfrac{3}{4}x^8$

Multiply the term by the exponent (power) (i.e. 8) and subtract 1 from the exponent. This gives:

$\dfrac{dy}{dx} = 8 \times \dfrac{3}{4}x^{8-1}$

$$= \frac{24}{4}x^7$$

$$\frac{dy}{dx} = 6x^7$$

(c) $y = 3\sqrt{x}$

Since $\sqrt{x} = x^{\frac{1}{2}}$, we can rewrite the expression as:

$$y = 3x^{\frac{1}{2}}$$

$$\frac{dy}{dx} = \frac{1}{2} \times 3x^{\frac{1}{2}-1}$$

$$= \frac{3}{2}x^{-\frac{1}{2}}$$

$$= \frac{3}{2} \times \frac{1}{x^{\frac{1}{2}}} \qquad \text{(Note that } x^{-\frac{1}{2}} = \frac{1}{x^{\frac{1}{2}}} \text{ from indices)}$$

$$\frac{dy}{dx} = \frac{3}{2x^{\frac{1}{2}}}$$

$$\frac{dy}{dx} = \frac{3}{2\sqrt{x}} \qquad \text{(Since } x^{\frac{1}{2}} = \sqrt{x} \text{)}$$

(d) $y = \dfrac{10}{\sqrt[5]{x^3}}$

Expressing the root in fractional form gives:

$$y = \frac{10}{(x^3)^{\frac{1}{5}}}$$

$$y = \frac{10}{x^{\frac{3}{5}}} \qquad \text{(The two exponents (powers) 3 and } \frac{1}{5} \text{ have been multiplied)}$$

Taking the denominator to the numerator changes the sign of its exponent as follows:

$$y = 10x^{-\frac{3}{5}}$$

Hence, $\dfrac{dy}{dx} = -\dfrac{3}{5} \times 10x^{-\frac{3}{5}-1}$

$$= -\frac{30}{5}x^{-\frac{8}{5}}$$

$$= -6x^{-\frac{8}{5}}$$

$$= -6 \times \frac{1}{x^{\frac{8}{5}}} \qquad \text{(Note that the inverse of a term changes the sign of its exponent)}$$

$$\frac{dy}{dx} = \frac{-6}{x^{\frac{8}{5}}}$$

Or, $\dfrac{dy}{dx} = \dfrac{-6}{\sqrt[5]{x^8}}$

(e) $y = \dfrac{1}{x^{\frac{1}{4}}}$

This can be expressed as:

$$y = x^{-\frac{1}{4}}$$

$$\frac{dy}{dx} = -\frac{1}{4} x^{-\frac{1}{4} - 1}$$

$$= -\frac{1}{4} x^{-\frac{5}{4}}$$

$$= -\frac{1}{4} \times \frac{1}{x^{\frac{5}{4}}}$$

$$\frac{dy}{dx} = -\frac{1}{4 x^{\frac{5}{4}}}$$ (Note that the inverse of a term changes the sign of its exponent)

Or, $\dfrac{dy}{dx} = \dfrac{1}{4 \sqrt[4]{x^5}}$

2. Find the derivative of each of the following:

(a) $5x^3 - 7x^2 - 3x + 8$

(b) $\dfrac{3}{5}x^5 + 2x^3 - x$

(c) $\dfrac{2x^4 - 5x^3 - 4x^2 + 3}{x^2}$

(d) $\sqrt{x} + \dfrac{1}{\sqrt{x}}$

<u>Solutions</u>

(a) Let the expression be $y = 5x^3 - 7x^2 - 3x + 8$

Hence, $\dfrac{dy}{dx} = \dfrac{d(5x^3)}{dx} - \dfrac{d(7x^2)}{dx} - \dfrac{d(3x)}{dx} + \dfrac{d(8)}{dx}$

This means that each part should be differentiated separately.

Hence, $\dfrac{dy}{dx} = (3 \times 5x^{3-1}) - (2 \times 7x^{2-1}) - (1 \times 3x^{1-1}) + 0$

$\dfrac{dy}{dx} = 15x^2 - 14x - 3$ (Note that $x^0 = 1$)

Note that the derivative of a constant is zero as shown by the derivative of 8

(b) $y = \dfrac{3}{5}x^5 + 2x^3 - x$

$\quad \dfrac{dy}{dx} = \left(5 \times \dfrac{3}{5}x^{5-1}\right) + \left(3 \times 2x^{3-1}\right) - \left(1 \times x^{1-1}\right)$

$\quad\quad = 3x^4 + 6x^2 - 1$

(c) $y = \dfrac{2x^4 - 5x^3 - 4x^2 + 3}{x^2}$

Dividing each term in the numerator by the denominator in order to separate the expression into it different fractions gives:

$\quad y = \dfrac{2x^4}{x^2} - \dfrac{5x^3}{x^2} - \dfrac{4x^2}{x^2} + \dfrac{3}{x^2}$

$\quad\quad = 2x^2 - 5x - 4 + \dfrac{3}{x^2}$

$\quad y = 2x^2 - 5x - 4 + 3x^{-2}$

$\quad \dfrac{dy}{dx} = (2 \times 2x) - 5 - 0 + \left(-2 \times 3x^{-2-1}\right)$

$\quad\quad = 4x - 5 - 6x^{-3}$

$\quad \dfrac{dy}{dx} = 4x - 5 - \dfrac{6}{x^2}$

(d) $y = \sqrt{x} + \dfrac{1}{\sqrt{x}}$

$\quad = x^{\frac{1}{2}} + \dfrac{1}{x^{\frac{1}{2}}}$

$\quad y = x^{\frac{1}{2}} + x^{-\frac{1}{2}}$

$\quad \dfrac{dy}{dx} = \dfrac{1}{2}x^{-\frac{1}{2}} - \dfrac{1}{2}x^{-\frac{3}{2}}$ $\quad$ (Note that $-\dfrac{1}{2} - 1 = -\dfrac{3}{2}$)

$\quad\quad = \dfrac{1}{2} \times \dfrac{1}{x^{\frac{1}{2}}} - \dfrac{1}{2} \times \dfrac{1}{x^{\frac{3}{2}}}$

$\quad\quad = \dfrac{1}{2x^{\frac{1}{2}}} - \dfrac{1}{2x^{\frac{3}{2}}}$

$\quad \dfrac{dy}{dx} = \dfrac{1}{2\sqrt{x}} - \dfrac{1}{2\sqrt{x^3}}$

3. If $y = (5x - 2)^3$, find $\dfrac{dy}{dx}$

<u>Solution</u>

$\quad y = (5x - 2)^3$ $\quad$ (This is a composite function)

165

Let us take u = $5x - 2$

If $5x - 2$ is replaced with U, then the question (i.e. y = $(5x - 2)^3$) becomes:

$$y = u^3$$

Hence, $\frac{dy}{du} = 3u^2$

Since u = $5x - 2$

then $\frac{du}{dx} = 5$

Therefore, $\frac{dy}{dx} = \frac{dy}{du} \times \frac{du}{dx}$ (Chain rule)

$$= 3u^2 \times 5$$

$$= 15u^2$$

Now substitute $5x - 2$ for u to obtain $\frac{dy}{dx}$ as follows:

$$\frac{dy}{dx} = 15(5x - 2)^2$$

4. If y = $\frac{1}{(5x^2 - 1)^4}$ find $\frac{dy}{dx}$

Solution

$$y = \frac{1}{(5x^2 - 1)^4}$$ (This is a composite function)

It can also be represented as follows:

$$y = (5x^2 - 1)^{-4}$$ (Its inverse changes the sign of its exponent)

Now, let us take u = $5x^2 - 1$

Hence, y = u^{-4} (When $5x^2 - 1$ is replaced with u in the original question)

Therefore, $\frac{dy}{du} = -4u^{-5}$

Since u = $5x^2 - 1$

then, $\frac{du}{dx} = 10x$

Therefore, $\frac{dy}{dx} = \frac{dy}{du} \times \frac{du}{dx}$ (Chain rule)

$$= -4u^{-5} \times 10x$$

$$\frac{dy}{dx} = -40xu^{-5}$$

Now substitute $5x^2 - 1$ for u. This gives:

$$\frac{dy}{dx} = -40x(5x^2 - 1)^{-5}$$

Or, $\frac{dy}{dx} = \frac{-40x}{(5x^2 - 1)^5}$ (Note the change in the sign of the exponent as it becomes

denominator)

5. If $y = (2x^3 + 7x)^{\frac{1}{2}}$ find $\dfrac{dy}{dx}$

<u>Solution</u>

$$y = (2x^3 + 7x)^{\frac{1}{2}}$$

Let $u = 2x^3 + 7x$

Hence, $y = u^{\frac{1}{2}}$

$$\frac{dy}{du} = \frac{1}{2}u^{-\frac{1}{2}}$$

Also, $u = 2x^3 + 7x$

$$\frac{du}{dx} = 6x^2 + 7$$

Therefore, $\dfrac{dy}{dx} = \dfrac{dy}{du} \times \dfrac{du}{dx}$

$$= \frac{1}{2}u^{-\frac{1}{2}} \times 6x^2 + 7$$

$$= \frac{6x^2 + 7}{2}u^{-\frac{1}{2}}$$

$$= \frac{6x^2 + 7}{2} \times \frac{1}{u^{1/2}}$$

$$\frac{dy}{dx} = \frac{6x^2 + 7}{2u^{1/2}}$$

Now, replace u with $2x^3 + 7x$. This gives:

$$\frac{dy}{dx} = \frac{6x^2 + 7}{2(2x^3 + 7x)^{1/2}}$$

Or, $\dfrac{dy}{dx} = \dfrac{6x^2 + 7}{2\sqrt{2x^3 + 7x}}$ (Note that $(2x^3 + 7x)^{\frac{1}{2}} = \sqrt{2x^3 + 7x}$)

6. Find the derivative of $3x^2 - x + 9)^4$

<u>Solution</u>

$$y = (3x^2 - x + 9)^4$$

Let $u = 3x^2 - x + 9$

Hence, $y = u^4$

$$\frac{dy}{du} = 4u^3$$

$$\frac{du}{dx} = 6x - 1$$

Therefore, $\dfrac{dy}{dx} = \dfrac{dy}{du} \times \dfrac{du}{dx}$

$$= 4u^3 \times 6x - 1$$

$$= 4(6x - 1)u^3$$

$$= (24x - 4)u^3$$

$$\frac{dy}{dx} = (24x - 4)(3x^2 - x + 9)^3 \quad \text{(When u is replaced with } 3x^2 - x + 9)$$

7. Find the derivative of $\left(x - \dfrac{5}{x}\right)^4$

<u>Solution</u>

$$y = \left(x - \frac{5}{x}\right)^4$$

Let $u = x - \dfrac{5}{x}$

Therefore, $y = u^4$

$$\frac{dy}{du} = 4u^3$$

$$\frac{du}{dx} = \frac{d(x)}{dx} - \frac{d(5x^{-1})}{dx} \quad \left(\text{Note that } = \frac{5}{x} = 5x^{-1}\right)$$

$$= 1 - (-1)5x^{-2}$$

$$= 1 + 5x^{-2}$$

$$\frac{du}{dx} = 1 + \frac{5}{x^2}$$

Therefore, $\dfrac{dy}{dx} = \dfrac{dy}{du} \times \dfrac{du}{dx}$

$$= 4u^3 \times \left(1 + \frac{5}{x^2}\right)$$

$$= \left(4 + \frac{20}{x^2}\right)u^3$$

$$\frac{dy}{dx} = \left(4 + \frac{20}{x^2}\right)\left(x - \frac{5}{x}\right)^3 \quad \left(\text{When u is replaced with } \left(x - \frac{5}{x}\right)\right)$$

8. Differentiate with respect to x: $\dfrac{1}{2x^5 - 3x + 1}$

<u>Solution</u>

Let $y = \dfrac{1}{2x^5 - 3x + 1}$

Or, $y = (2x^5 - 3x + 1)^{-1} \quad \left(\text{Recall from indices that } \dfrac{1}{a} = a^{-1}\right)$

Let us take $u = 2x^5 - 3x + 1$

Therefore, $y = u^{-1}$

$$\frac{dy}{du} = -1u^{-2}$$

$$\frac{du}{dx} = 10x^4 - 3$$

Therefore, $\dfrac{dy}{dx} = \dfrac{dy}{du} \times \dfrac{du}{dx}$

$$= -1u^{-2} \times 10x^4 - 3$$

$$= -1(10x^4 - 3)u^{-2}$$

$$= (-10x^4 + 3)u^{-2}$$

$$= \frac{3 - 10x^4}{u^2}$$

$$\frac{dy}{dx} = \frac{3 - 10x^4}{(2x^5 - 3x + 1)^2}$$

9. Differentiate with respect to x: $\sqrt{7 - 5x^3}$

<u>Solution</u>

Let $y = \sqrt{7 - 5x^3}$

Or, $y = (7 - 5x^3)^{\frac{1}{2}}$ (Recall from indices that $\sqrt{a} = a^{\frac{1}{2}}$)

Let $u = 7 - 5x^3$

Hence, $y = u^{\frac{1}{2}}$

$$\frac{dy}{du} = \frac{1}{2}u^{-\frac{1}{2}}$$

$$\frac{du}{dx} = -15x^2$$

Therefore, $\dfrac{dy}{dx} = \dfrac{dy}{du} \times \dfrac{du}{dx}$

$$= \frac{1}{2}u^{-\frac{1}{2}} \times (-15x^2)$$

$$= -\frac{15}{2}x^2 \left(u^{-\frac{1}{2}}\right)$$

$$= -\frac{15x^2}{2u^{\frac{1}{2}}}$$

$$= -\frac{15x^2}{2\sqrt{u}}$$

$$\frac{dy}{dx} = -\frac{15x^2}{2\sqrt{7 - 5x^3}}$$

10. Find $\dfrac{dy}{dx}$ if $y = \dfrac{1}{\sqrt{2x^3 - 5}}$

<u>Solution</u>

$$y = \frac{1}{\sqrt{2x^3 - 5}}$$

Let $u = 2x^3 - 5$

hence, $y = \dfrac{1}{\sqrt{u}}$

$$y = \frac{1}{u^{\frac{1}{2}}}$$

$$y = u^{-\frac{1}{2}}$$

$$\frac{dy}{du} = -\frac{1}{2}u^{-\frac{3}{2}}$$ (Note that $-\frac{1}{2} - 1 = -\frac{3}{2}$)

$$\frac{du}{dx} = 6x^2$$

Therefore, $\dfrac{dy}{dx} = \dfrac{dy}{du} \times \dfrac{du}{dx}$

$$= -\frac{1}{2}u^{-\frac{3}{2}} \times (6x^2)$$

$$= -\frac{6}{2}x^2 \, (u^{-\frac{3}{2}})$$

$$= -3x^2(u^{-\frac{3}{2}})$$

$$= -\frac{3x^2}{u^{\frac{3}{2}}}$$

$$= -\frac{3x^2}{\sqrt{u^3}}$$

$$\frac{dy}{dx} = -\frac{3x^2}{\sqrt{(2x^3 - 5)^3}}$$

Exercise 10

1. Find the derivatives of the following:

(a) $y = 8x^5$

(b) $y = \dfrac{2}{5}x^5$

(c) $y = \sqrt[3]{x}$

(d) $y = 7\sqrt[7]{x}$

(e) $y = \dfrac{1}{\sqrt[8]{x^5}}$

(f) $y = \dfrac{2}{x^{\frac{5}{2}}}$

2. Find the derivative of each of the following:

(a) $2x^5 - 3x^4 - 4x^3 + 5x^2 - 6x + 7$

(b) $x^7 + 2x^4 - \dfrac{3}{x}$

(c) $\dfrac{3x^9 - x^7 - 5x^4 + 2x^2 - 1}{x^3}$

(d) $5(\sqrt[4]{x}) + \dfrac{5}{\sqrt[3]{2x}}$

3. If $y = (2x - 5)^4$, find $\dfrac{dy}{dx}$

4. If $y = \dfrac{3}{(x^3 - 7)^2}$ find $\dfrac{dy}{dx}$

5. If $y = (2x^3 + 7x)^{\frac{1}{2}}$ find $\dfrac{dy}{dx}$

6. Find the derivative of $(7x^3 - x^2 + 3)^5$

7. Find the derivative of $\left(3x - \dfrac{2}{3x}\right)^3$

8. Differentiate with respect to x: $-\dfrac{9}{3x^2 - x - 10}$

9. Differentiate with respect to x: $\sqrt{1 - 2x^4}$

10. Find $\dfrac{dy}{dx}$ if $y = \dfrac{1}{\sqrt[3]{5x^3 - 1}}$

11. Find the derivative of $(x^5 - 3)^9$

12. Find the derivative of $\left(x - \dfrac{1}{5x}\right)^2$

13. Differentiate with respect to x: $\dfrac{2}{x^3 - x - \frac{1}{x}}$

14. Differentiate with respect to x: $\sqrt{5x - x^2}$

15. Find $\dfrac{dy}{dx}$ if $y = \dfrac{1}{\sqrt[5]{x^3 + 2}}$

CHAPTER 11
PRODUCT RULE OF DERIVATIVE

If $y = uv$ where u and v are functions of x, then:

$$\frac{dy}{dx} = u\frac{dv}{dx} + v\frac{du}{dx}$$

This is called the product rule of differentiation.

Similarly, If $y = uvw$ where u, v and w are functions of x, then:

$$\frac{dy}{dx} = \frac{du}{dx}vw + \frac{dv}{dx}uw + \frac{dw}{dx}uv$$

Examples

1. If $f(x) = (x - 3)(x + 4)$, find $f'(x)$

<u>Solution</u>

$$f(x) = (x - 3)(x + 4)$$

This is a product of two functions of x.

Let $u = x - 3$

and $v = x + 4$

Hence, $\dfrac{du}{dx} = 1$

$\dfrac{dv}{dx} = 1$

Therefore, the derivative of $f(x)$ is:

$$f'(x) = u\frac{dv}{dx} + v\frac{du}{dx}$$
$$= (x - 3) \times 1 + (x + 4) \times 1$$
$$= x - 3 + x + 4$$
$$f'(x) = 2x + 1$$

Note that another way to differentiate the function in example 1 above is to expand the bracket and differentiate directly.

2. Find the derivative of $y = (4x^2 + 1)(x^2 - 3)$

<u>Solution</u>

$$y = (4x^2 + 1)(x^2 - 3)$$

Let $u = 4x^2 + 1$

and $v = (x^2 - 3)$

Hence, $\dfrac{du}{dx} = 8x$

$\dfrac{dv}{dx} = 2x$

$$\frac{dy}{dx} = u\frac{dv}{dx} + v\frac{du}{dx}$$

$$= (4x^2 + 1)2x + (x^2 - 3)8x$$

$$= 8x^3 + 2x + 8x^3 - 24x$$

$$\frac{dy}{dx} = 16x^3 - 22x$$

3. If $y = x^2(1 + 2x)^{\frac{1}{2}}$, find $\frac{dy}{dx}$

Solution

$$y = x^2(1 + 2x)^{\frac{1}{2}}$$

$$u = x^2$$

and $v = (1 + 2x)^{\frac{1}{2}}$

Hence, $\frac{du}{dx} = 2x$

$$\frac{dv}{dx} = \frac{1}{2} \times 2 \times (1 + 2x)^{\frac{1}{2} - 1} \qquad \text{(Use of chain rule)}$$

$$= (1 + 2x)^{-\frac{1}{2}}$$

$$\frac{dv}{dx} = \frac{1}{(1+2x)^{\frac{1}{2}}}$$

Hence, $\frac{dy}{dx} = u\frac{dv}{dx} + v\frac{du}{dx}$

$$= x^2 \frac{1}{(1+2x)^{\frac{1}{2}}} + (1 + 2x)^{\frac{1}{2}} \times 2x$$

$$= \frac{x^2}{(1+2x)^{\frac{1}{2}}} + 2x(1 + 2x)^{\frac{1}{2}}$$

let us now simplify by using $(1 + 2x)^{\frac{1}{2}}$ as the LCM as follows:

$$= \frac{x^2 + 2x\,(1+2x)^{\frac{1}{2}}(1+2x)^{\frac{1}{2}}}{(1+2x)^{\frac{1}{2}}}$$

$$= \frac{x^2 + 2x\,(1+2x)}{(1+2x)^{\frac{1}{2}}} \qquad \text{(Note that } (1 + 2x)^{\frac{1}{2}} \times (1 + 2x)^{\frac{1}{2}} = (1 + 2x)^{\frac{1}{2} + \frac{1}{2}} = 1 + 2x\text{)}$$

$$= \frac{x^2 + 2x + 4x^2}{(1+2x)^{\frac{1}{2}}}$$

$$\frac{dy}{dx} = \frac{5x^2 + 2x}{\sqrt{1+2x}}$$

4. Find the derivative of $(2x + 3)^3(4x^2 - 1)^2$

Solution

$$y = (2x + 3)^3(4x^2 - 1)^2$$

173

Let $u = (2x + 3)^3$

and $v = (4x^2 - 1)^2$

$\dfrac{du}{dx} = 3(2x + 3)^{3-1} \times 2$ (Use of chain rule. Also note that 2 is from the derivative of $2x + 3$)

$= 6(2x + 3)^2$

$\dfrac{dv}{dx} = 2(4x^2 - 1)^{2-1} \times 8x$ (Note that $8x$ is from the derivative of $4x^2 - 1$)

$= 16x(4x^2 - 1)$

Hence, $\dfrac{dy}{dx} = u\dfrac{dv}{dx} + v\dfrac{du}{dx}$

$= (2x + 3)^3 \times 16x(4x^2 - 1) + (4x^2 - 1)^2 \times 6(2x + 3)^2$

Let us factorize the expression by taking out $(2x + 3)^2$ and $(4x^2 - 1)$ which are the common terms as follows:

$\dfrac{dy}{dx} = (2x + 3)^2(4x^2 - 1)[(2x + 3)16x + (4x^2 - 1)6]$

$= (2x + 3)^2(4x^2 - 1)(32x^2 + 48x + 24x^2 - 6)$

$= (2x + 3)^2(4x^2 - 1)(56x^2 + 48x - 6)$

$= (2x + 3)^2(4x^2 - 1)\,2(28x^2 + 24x - 3)$

$\dfrac{dy}{dx} = 2(2x + 3)^2(4x^2 - 1)(28x^2 + 24x - 3)$

5. Differentiate: $y = x(x + 1)(x^2 - 4)$

Solution

$y = x(x + 1)(x^2 - 4)$

Expanding the first two brackets gives:

$y = (x^2 + x)(x^2 - 4)$

Hence, $u = (x^2 + x)$

and $v = (x^2 - 4)$

Hence, $\dfrac{du}{dx} = 2x + 1$

$\dfrac{dv}{dx} = 2x$

Therefore, $\dfrac{dy}{dx} = u\dfrac{dv}{dx} + v\dfrac{du}{dx}$

$= (x^2 + x)2x + (x^2 - 4)(2x + 1)$

$= 2x^3 + 2x^2 + 2x^3 + x^2 - 8x - 4$

$\dfrac{dy}{dx} = 4x^3 + 3x^2 - 8x - 4$

6. Find the derivative of $(x^2 + 3x - 2)^2 \sqrt{x}$

Solution

$y = (x^2 + 3x - 2)^2 \sqrt{x}$

$$u = (x^2 + 3x - 2)^2$$

$$v = \sqrt{x}$$

Or, $v = x^{\frac{1}{2}}$

$$\frac{du}{dx} = 2(x^2 + 3x - 2)^{2-1} \times 2x + 3 \quad \text{(Note that } 2x + 3 \text{ is from the derivative of } x^2 + 3x - 2)$$

$$= (4x + 6)(x^2 + 3x - 2)$$

$$\frac{dv}{dx} = \frac{1}{2} x^{-\frac{1}{2}} \quad \text{(Note that } 8x \text{ is from the derivative of } 4x^2 - 1)$$

$$= \frac{1}{2x^{\frac{1}{2}}}$$

$$\frac{dv}{dx} = \frac{1}{2\sqrt{x}}$$

Hence, $\dfrac{dy}{dx} = u\dfrac{dv}{dx} + v\dfrac{du}{dx}$

$$= (x^2 + 3x - 2)^2 \frac{1}{2\sqrt{x}} + \sqrt{x}(4x + 6)(x^2 + 3x - 2)$$

$$= (x^2 + 3x - 2)^2 \frac{1}{2\sqrt{x}} + 2\sqrt{x}(2x + 3)(x^2 + 3x - 2)$$

Factorize the expression by taking out ($x^2 + 3x - 2$) which is the common factor. This gives:

$$\frac{dy}{dx} = (x^2 + 3x - 2)\left[\frac{x^2 + 3x - 2}{2\sqrt{x}} + 2\sqrt{x}(2x + 3)\right]$$

Simplifying the part in the bracket by taking $2\sqrt{x}$ as the LCM gives:

$$\frac{dy}{dx} = (x^2 + 3x - 2)\left[\frac{x^2 + 3x - 2 + 4x(2x + 3)}{2\sqrt{x}}\right] \quad \text{(Note that } 2\sqrt{x} \times 2\sqrt{x} = 4x)$$

$$= (x^2 + 3x - 2)\left[\frac{x^2 + 3x - 2 + 8x^2 + 12x}{2\sqrt{x}}\right]$$

$$\frac{dy}{dx} = \frac{(x^2 + 3x - 2)(9x^2 + 15x - 2)}{2\sqrt{x}}$$

7. Find the derivative of $(1 + x)(5x - 2)^{\frac{3}{2}}$

Solution

$$y = (1 + x)(5x - 2)^{\frac{3}{2}}$$

$$u = (1 + x)$$

$$v = (5x - 2)^{\frac{3}{2}}$$

$$\frac{du}{dx} = 1$$

$$\frac{dv}{dx} = \frac{3}{2}(5x - 2)^{\frac{3}{2} - 1} \times 5 \quad \text{(Note that the derivative of } 5x - 2 \text{ is 5)}$$

$$= \frac{15}{2}(5x - 2)^{\frac{1}{2}}$$

Hence, $\dfrac{dy}{dx} = u\dfrac{dv}{dx} + v\dfrac{du}{dx}$

175

$$= (1 + x) \times \frac{15}{2}(5x - 2)^{\frac{1}{2}} + (5x - 2)^{\frac{3}{2}} \times 1$$

$$= \frac{15}{2}(1 + x)(5x - 2)^{\frac{1}{2}} + (5x - 2)^{\frac{3}{2}}$$

Factorize by taking out $(5x - 2)^{\frac{1}{2}}$ (i.e. the lower exponent) which is the common factor gives:

$$\frac{dy}{dx} = (5x - 2)^{\frac{1}{2}}\left[\frac{15}{2}(1 + x) + 5x - 2\right] \quad \left(\text{Note that } \frac{(5x-2)^{\frac{3}{2}}}{(5x-2)^{\frac{1}{2}}} = (5x - 2)^{\frac{3}{2} - \frac{1}{2}} = 5x - 2\right)$$

$$= (5x - 2)^{\frac{1}{2}}\left[\frac{15}{2} + \frac{15x}{2} + 5x - 2\right]$$

$$= (5x - 2)^{\frac{1}{2}}\left[\frac{25x}{2} + \frac{11}{2}\right]$$

$$\frac{dy}{dx} = \sqrt{5x - 2}\left[\frac{25x + 11}{2}\right]$$

8. If $y = (1 + x)(2 - 3x)(2x - 1)$, find $\frac{dy}{dx}$ by using product rule.

Solution

$$y = (1 + x)(2 - 3x)(2x - 1)$$

This is a product of three expressions, u, v and w.

Hence, u = $(1 + x)$

v = $(2 - 3x)$

and   w = $(2x - 1)$

Therefore, $\dfrac{du}{dx} = 1$

$\dfrac{dv}{dx} = -3$

$\dfrac{dw}{dx} = 2$

Hence the formula for product rule of three terms is given by:

$$\frac{dy}{dx} = \frac{du}{dx}vw + \frac{dv}{dx}uw + \frac{dw}{dx}uv$$

$$= 1(2 - 3x)(2x - 1) + (-3)(1 + x)(2x - 1) + 2(1 + x)(2 - 3x)$$

$$= 4x - 2 - 6x^2 + 3x + (-3 - 3x)(2x - 1) + (2 + 2x)(2 - 3x)$$

$$= 7x - 2 - 6x^2 - 6x + 3 - 6x^2 + 3x + 4 - 6x + 4x - 6x^2$$

$$= -2 + 3 + 4 + 7x - 6x + 3x - 6x + 4x - 6x^2 - 6x^2 - 6x^2$$

$$\frac{dy}{dx} = 5 + 2x - 18x^2$$

9. Differentiate with respect to x: $(x^2 - 3x + 5)(2x - 7)$

Solution

$$y = (x^2 - 3x + 5)(2x - 7)$$

Let us differentiate this product by applying product rule but without the use u and v. This is done as follows:

$$\frac{dy}{dx} = (x^2 - 3x + 5)\frac{d(2x-7)}{dx} + (2x - 7)\frac{d(x^2 - 3x + 5)}{dx}$$

$$= (x^2 - 3x + 5)2 + (2x - 7)(2x - 3)$$

$$= 2x^2 - 6x + 10 + 4x^2 - 6x - 14x + 21$$

$$\frac{dy}{dx} = 6x^2 - 26x + 31$$

10. If $y = (5x^2 - 3)(2 + \frac{3}{x})$, find $\frac{dy}{dx}$

Solution

$$y = (5x^2 - 3)(2 + \frac{3}{x})$$

$$u = (5x^2 - 3)$$

$$v = (2 + \frac{3}{x})$$

$$\frac{du}{dx} = 10x$$

$$\frac{dv}{dx} = \frac{d(3x^{-1})}{dx} \qquad \text{(Note that } \frac{3}{x} = 3x^{-1})$$

$$= -3x^{-2}$$

$$\frac{dv}{dx} = \frac{-3}{x^2}$$

Hence, $\frac{dy}{dx} = u\frac{dv}{dx} + v\frac{du}{dx}$

$$= (5x^2 - 3)\left(\frac{-3}{x^2}\right) + (2 + \frac{3}{x})10x$$

$$= -15 + \frac{9}{x^2} + 20x + 30$$

$$\frac{dy}{dx} = 15 + 20x + \frac{9}{x^2}$$

Exercise 11

1. If $f(x) = (2x - 1)(3x + 1)$, find $f'(x)$
2. Find the derivative of $y = (3x^2 - 5)(x^2 + 10)$
3. If $y = 5x(3 + x)^{\frac{1}{2}}$ find $\frac{dy}{dx}$
4. Find the derivative of $(x + 5)(x^2 - 7)^3$
5. Differentiate: $y = 2x(3x + 2)(2x^2 - 5)$
6. Find the derivative of $(3x^2 - 1)^3\sqrt{2x}$

7. Find the derivative of $(9 - x)(x + 3)^{\frac{3}{4}}$

8. If $y = (2 - x)(5 - 3x^2)(x + 3)$, find $\dfrac{dy}{dx}$ by using product rule.

9. Differentiate with respect to x: $(3x^4 - x^2 + 2x)(x - 1)$

10. If $y = (3x^2 - x)(1 + \dfrac{1}{2x})$, find $\dfrac{dy}{dx}$

11. Find the derivative of $x^2(\sqrt{x^5})$

12. Find the derivative of $x^4(2x - 11)^{\frac{2}{3}}$

13. If $y = (7 + x)(1 - x^2)(5x^3 + 1)$, find $\dfrac{dy}{dx}$.

14. Differentiate with respect to x: $(3x^4 - x^3 + x^2 + 2x - 3)(5x + 4)$

15. If $y = (x^3 - 3x + 5)\left(\dfrac{1}{x^5}\right)$, find $\dfrac{dy}{dx}$

178

CHAPTER 12
QUOTIENT RULE OF DERIVATIVE

If $y = \dfrac{u}{v}$ where u and v are functions of x, then:

$$\frac{dy}{dx} = \frac{v\frac{du}{dx} - u\frac{dv}{dx}}{v^2}$$

This is called the quotient rule of differentiation.

Examples

1. If $y = \dfrac{3x^2 - 8x + 5}{5x - 2}$ find $\dfrac{dy}{dx}$

Solution

$$y = \frac{3x^2 - 8x + 5}{5x - 2}$$

This is of the form $y = \dfrac{u}{v}$. Therefore, we are going to apply product rule.

Let $u = 3x^2 - 8x + 5$

and $v = 5x - 2$

Hence, $\dfrac{du}{dx} = 6x - 8$

$\dfrac{dv}{dx} = 5$

Therefore, $\dfrac{dy}{dx} = \dfrac{v\frac{du}{dx} - u\frac{dv}{dx}}{v^2}$

$$= \frac{(5x - 2)(6x - 8) - (3x^2 - 8x + 5)(5)}{(5x - 2)^2}$$

$$= \frac{30x^2 - 40x - 12x + 16 - 15x^2 - 40x - 25}{(5x - 2)^2}$$

$$\frac{dy}{dx} = \frac{15x^2 - 12x - 9}{(5x - 2)^2}$$

2. Differentiate with respect to x, the function: $\dfrac{3x^2 - 2x}{x + 5}$

Solution

$$y = \frac{3x^2 - 2x}{x + 5}$$

Let $u = 3x^2 - 2x$

and $v = x + 5$

Hence, $\dfrac{du}{dx} = 6x - 2$

$\dfrac{dv}{dx} = 1$

Therefore, $\dfrac{dy}{dx} = \dfrac{v\frac{du}{dx} - u\frac{dv}{dx}}{v^2}$

$= \dfrac{(x+5)(6x-2) - (3x^2 - 2x)(1)}{(x+5)^2}$

$= \dfrac{6x^2 - 2x + 30x - 10 - 3x^2 + 2x}{(x+5)^2}$

$\dfrac{dy}{dx} = \dfrac{3x^2 + 30x - 10}{(x+5)^2}$

3. Find the differential coefficient of $y = \dfrac{-3}{x^2 + 5}$

Solution

$y = \dfrac{-3}{x^2 + 5}$

$u = -3$

and $v = x^2 + 5$

Hence, $\dfrac{du}{dx} = 0$ (The derivative of a constant is zero)

$\dfrac{dv}{dx} = 2x$

Therefore, $\dfrac{dy}{dx} = \dfrac{v\frac{du}{dx} - u\frac{dv}{dx}}{v^2}$

$= \dfrac{(x^2 + 5) \times 0 - (-3)2x}{(x^2 + 5)^2}$

$= \dfrac{0 - 6x}{(x^2 + 5)^2}$

$\dfrac{dy}{dx} = \dfrac{-6x}{(x^2 + 5)^2}$

4. Differentiate: $y = \dfrac{\sqrt{3-x}}{\sqrt{3+x}}$

Solution

$y = \dfrac{\sqrt{3-x}}{\sqrt{3+x}}$

$u = \sqrt{3-x}$

$= (3-x)^{\frac{1}{2}}$

and $v = \sqrt{3+x}$

$= (3+x)^{\frac{1}{2}}$

Hence, $\dfrac{du}{dx} = \dfrac{1}{2}(3-x)^{\frac{1}{2}-1} \times -1$ (Note that the derivative of $3 - x$ is -1)

$= -\dfrac{1}{2}(3-x)^{-\frac{1}{2}}$

$$\frac{du}{dx} = \frac{-(3-x)^{-\frac{1}{2}}}{2}$$

$$\frac{dv}{dx} = \frac{1}{2}(3+x)^{\frac{1}{2}-1} \times 1 \quad \text{(Note that the derivative of } 3-x \text{ is } -1)$$

$$= \frac{1}{2}(3+x)^{-\frac{1}{2}}$$

$$\frac{dv}{dx} = \frac{(3+x)^{-\frac{1}{2}}}{2}$$

Therefore, $\dfrac{dy}{dx} = \dfrac{v\frac{du}{dx} - u\frac{dv}{dx}}{v^2}$

$$= \frac{(3+x)^{\frac{1}{2}}\left(\frac{-(3-x)^{-\frac{1}{2}}}{2}\right) - (3-x)^{\frac{1}{2}}\left(\frac{(3+x)^{-\frac{1}{2}}}{2}\right)}{((3+x)^{\frac{1}{2}})^2}$$

Taking out the common terms which are terms with positive exponent in order to factorize the expression gives:

$$\frac{dy}{dx} = \frac{\left((3+x)^{\frac{1}{2}}\right)\left((3-x)^{\frac{1}{2}}\right)\left[\frac{-(3-x)^{-1}}{2} - \frac{(3+x)^{-1}}{2}\right]}{3+x}$$

Note that in order to obtain the terms in the square bracket, we subtracted the exponents of the factors from the exponents of the original term. For example, $(3+x)^{\frac{1}{2}}\left(\frac{-(3-x)^{-\frac{1}{2}}}{2}\right)$ divided by $\left((3+x)^{\frac{1}{2}}\right)\left((3-x)^{\frac{1}{2}}\right)$ gave $\frac{-(3-x)^{-1}}{2}$ since the equal terms canceled out and $-\frac{1}{2} - \frac{1}{2} = -1$, which gave the exponent of -1. Similarly, $(3-x)^{\frac{1}{2}}\left(\frac{(3+x)^{-\frac{1}{2}}}{2}\right)$ divided by $\left((3+x)^{\frac{1}{2}}\right)\left((3-x)^{\frac{1}{2}}\right)$ gave $\frac{(3+x)^{-1}}{2}$ since the equal terms canceled out and $-\frac{1}{2} - \frac{1}{2} = -1$, which gave the exponent of -1 as the terms in the square bracket.

Let us now continue with the solution by simplifying further as follows:

$$\frac{dy}{dx} = \frac{\left((3+x)^{\frac{1}{2}}\right)\left((3-x)^{\frac{1}{2}}\right)\left[\frac{-(3-x)^{-1}}{2} - \frac{(3+x)^{-1}}{2}\right]}{3+x}$$

$$= \frac{\left((3+x)^{\frac{1}{2}}\right)\left((3-x)^{\frac{1}{2}}\right)\left[\frac{-1}{2(3-x)} - \frac{1}{2(3+x)}\right]}{3+x}$$

$$= \frac{\left((3+x)^{\frac{1}{2}}\right)\left((3-x)^{\frac{1}{2}}\right)\left[\frac{-(3+x) - (3-x)}{2(3-x)(3+x)}\right]}{3+x}$$

181

$$= \frac{\left((3+x)^{\frac{1}{2}}\right)\left((3-x)^{\frac{1}{2}}\right)\left[\frac{-3-x-3+x}{2(3-x)(3+x)}\right]}{3+x}$$

$$= \frac{\left((3+x)^{\frac{1}{2}}\right)\left((3-x)^{\frac{1}{2}}\right)\left[\frac{-6}{2(3-x)(3+x)}\right]}{3+x}$$

$$= \frac{\left((3+x)^{\frac{1}{2}}\right)\left((3-x)^{\frac{1}{2}}\right)\left[\frac{-3}{(3-x)(3+x)}\right]}{3+x}$$

$$= \frac{-3\left((3+x)^{\frac{1}{2}}\right)\left((3-x)^{\frac{1}{2}}\right)}{(3-x)(3+x)(3+x)}$$

$$= \frac{-3\left((3+x)^{\frac{1}{2}}\right)\left((3-x)^{\frac{1}{2}}\right)}{(3+x)^2(3-x)}$$

$$= -3(3+x)^{\frac{1}{2}-2}(3-x)^{\frac{1}{2}-1} \quad \text{(Subtraction of exponents due to the division above)}$$

$$= -3(3+x)^{-\frac{3}{2}}(3-x)^{-\frac{1}{2}}$$

$$= \frac{-3}{(3+x)^{\frac{3}{2}}(3-x)^{\frac{1}{2}}}$$

$$= \frac{-3}{\sqrt{(3+x)^3(3-x)}}$$

5. Differentiate with respect to x: $y = \frac{(2x^2-3)^3}{x}$

<u>Solution</u>

$$y = \frac{(2x^2-3)^3}{x}$$
$$u = (2x^2-3)^3$$

and $v = x$

Hence, $\frac{du}{dx} = 3(2x^2-3)^{3-1} \times 4x$

$$= 12x(2x^2-3)^2$$

$$\frac{dv}{dx} = 1$$

Therefore, $\frac{dy}{dx} = \frac{v\frac{du}{dx} - u\frac{dv}{dx}}{v^2}$

$$= \frac{x[12x(2x^2-3)^2] - (2x^2-3)^3 \times 1}{x^2}$$

$$= \frac{12x^2(2x^2-3)^2 - (2x^2-3)^3}{x^2}$$

Factorize the expression by taking out $(2x^2-3)^2$ which has the lower exponent. This gives:

$$\frac{dy}{dx} = \frac{(2x^2-3)^2[12x^2-(2x^2-3)]}{x^2}$$

$$= \frac{(2x^2-3)^2(12x^2-2x^2+3)}{x^2}$$

$$\frac{dy}{dx} = \frac{(2x^2-3)^2(10x^2+3)}{x^2}$$

6. Find the derivative of $\dfrac{\sqrt{(1+2x^2)^3}}{x}$

Solution

$$y = \frac{\sqrt{(1+2x^2)^3}}{x}$$

$$u = \sqrt{(1+2x^2)^3}$$

$$= [(1+2x^2)^3]^{\frac{1}{2}} \qquad \text{(When the square root sign is removed, we use an exponent of } \tfrac{1}{2})$$

$$u = (1+2x^2)^{\frac{3}{2}} \qquad \text{(After multiplying the exponents)}$$

$$v = x$$

Hence, $\dfrac{du}{dx} = \dfrac{3}{2}(1+2x^2)^{\frac{3}{2}-1} \times 4x$ (Note that the derivative of $3-x$ is -1)

$$= 6x(1+2x^2)^{\frac{1}{2}}$$

$$\frac{dv}{dx} = 1$$

Therefore, $\dfrac{dy}{dx} = \dfrac{v\frac{du}{dx} - u\frac{dv}{dx}}{v^2}$

$$= \frac{x\left(6x(1+2x^2)^{\frac{1}{2}}\right) - (1+2x^2)^{\frac{3}{2}} \times 1}{x^2}$$

Take out $(1+2x^2)^{\frac{1}{2}}$ which has the lower exponent and factorize the expression. This gives:

$$\frac{dy}{dx} = \frac{(1+2x^2)^{\frac{1}{2}}[6x^2-(1+2x^2)]}{x^2} \qquad \text{(Note that } \frac{(1+2x^2)^{\frac{3}{2}}}{(1+2x^2)^{\frac{1}{2}}} \text{ gives } (1+2x^2) \text{ by subtracting exponents.}$$

$$= \frac{(1+2x^2)^{\frac{1}{2}}(6x^2-1-2x^2)}{x^2}$$

$$= \frac{(1+2x^2)^{\frac{1}{2}}(4x^2-1)}{x^2}$$

$$\frac{dy}{dx} = \frac{(\sqrt{1+2x^2})(4x^2-1)}{x^2}$$

7. If $y = \dfrac{(4x^3 - 3x^2 + x + 1)^{\frac{1}{2}}}{(x+1)^2}$

<u>Solution</u>

$$y = \dfrac{(4x^3 - 3x^2 + x + 1)^{\frac{1}{2}}}{(x+1)^2}$$

$u = (4x^3 - 3x^2 + x + 1)^{\frac{1}{2}}$　　　(After multiplying the exponents)

$v = (x + 1)^2$

Hence, $\dfrac{du}{dx} = \dfrac{1}{2}(4x^3 - 3x^2 + x + 1)^{\frac{1}{2} - 1} \times (12x^2 - 6x + 1)$

$$= \dfrac{(12x^2 - 6x + 1)(4x^3 - 3x^2 + x + 1)^{-\frac{1}{2}}}{2}$$

$\dfrac{dv}{dx} = 2(x + 1)^{2-1}$

$\qquad = 2(x + 1)$

Therefore, $\dfrac{dy}{dx} = \dfrac{v\dfrac{du}{dx} - u\dfrac{dv}{dx}}{v^2}$

$$= \dfrac{\dfrac{(x+1)^2(12x^2 - 6x + 1)(4x^3 - 3x^2 + x + 1)^{-\frac{1}{2}} - (4x^3 - 3x^2 + x + 1)^{\frac{1}{2}}[2(x+1)]}{2}}{[(x+1)^2]^2}$$

Take out the common terms with lower exponents [i.e. $(x + 1)$ and $(4x^3 - 3x^2 + x + 1)^{-\frac{1}{2}}$] and factorize the expression. This gives:

$$\dfrac{dy}{dx} = \dfrac{\dfrac{(x+1)(4x^3 - 3x^2 + x + 1)^{-\frac{1}{2}}[(x+1)(12x^2 - 6x + 1) - (4x^3 - 3x^2 + x + 1)2]}{2}}{[(x+1)^2]^2}$$

Remember to subtract the exponents of the factors from the exponents of the original expression when simplifying. Simplifying further, the expression above gives:

$$\dfrac{dy}{dx} = \dfrac{(x+1)(4x^3 - 3x^2 + x + 1)^{-\frac{1}{2}}[12x^3 - 6x^2 + x + 12x^2 - 6x + 1 - (8x^3 - 6x^2 + 2x + 2)]}{2(x+1)^4}$$

$$= \dfrac{(x+1)(4x^3 - 3x^2 + x + 1)^{-\frac{1}{2}}(12x^3 - 6x^2 + x + 12x^2 - 6x + 1 - 8x^3 + 6x^2 - 2x - 2)}{2(x+1)^4}$$

$$= \dfrac{(x+1)(4x^3 - 3x^2 + x + 1)^{-\frac{1}{2}}(4x^3 + 12x^2 - 7x - 1)}{2(x+1)^4}$$

$$= \dfrac{(4x^3 - 3x^2 + x + 1)^{-\frac{1}{2}}(4x^3 + 12x^2 - 7x - 1)}{2(x+1)^3}$$

Note that $(x + 1)$ cancels out from the numerator and denominator.

Therefore, $\dfrac{dy}{dx} = \dfrac{4x^3 + 12x^2 - 7x - 1}{2(4x^3 - 3x^2 + x + 1)^{\frac{1}{2}}(x+1)^3}$

8. Determine $\dfrac{d}{dx}\left(\dfrac{3 + 2x - x^2}{\sqrt{1+x}}\right)$

Solution

Let $y = \dfrac{3 + 2x - x^2}{\sqrt{1+x}}$

Hence, $u = 3 + 2x - x^2$

$\dfrac{du}{dx} = 2 - 2x$

$v = \sqrt{1 + x}$

$= (1 + x)^{\frac{1}{2}}$

$\dfrac{du}{dx} = \dfrac{1}{2}(1 + x)^{\frac{1}{2} - 1} \times 1$

$= \dfrac{1}{2}(1 + x)^{-\frac{1}{2}}$

Therefore, $\dfrac{dy}{dx} = \dfrac{v\frac{du}{dx} - u\frac{dv}{dx}}{v^2}$

$= \dfrac{(1 + x)^{\frac{1}{2}}(2 - 2x) - (3 + 2x - x^2)\frac{1}{2}(1 + x)^{-\frac{1}{2}}}{[(1 + x)^{\frac{1}{2}}]^2}$

Take out $(1 + x)^{-\frac{1}{2}}$ as the common factor since it has the lower exponent and factorize the expression. This gives:

$\dfrac{dy}{dx} = \dfrac{(1 + x)^{-\frac{1}{2}}\left[(1+x)(2 - 2x) - (3 + 2x - x^2)\frac{1}{2}\right]}{1 + x}$

$= \dfrac{(1 + x)^{-\frac{1}{2}}\left[\frac{2(1+x)(2 - 2x) - (3 + 2x - x^2)}{2}\right]}{1 + x}$

$= \dfrac{(1 + x)^{-\frac{1}{2}}[(2 + 2x)(2 - 2x) - 3 - 2x + x^2]}{2(1 + x)}$

$= \dfrac{(1 + x)^{-\frac{1}{2}}(4 - 4x + 4x - 4x^2 - 3 - 2x + x^2)}{2(1 + x)}$

$$= \frac{1-2x-3x^2}{2(1+x)^{\frac{1}{2}}(1+x)} \qquad \text{[When } (1+x)^{-\frac{1}{2}} \text{ is taken to the denominator it becomes } (1+x)^{\frac{1}{2}}\text{]}$$

$$\frac{dy}{dx} = \frac{1-2x-3x^2}{2(1+x)^{\frac{3}{2}}} \qquad \text{(The exponents of same terms have been added together, i.e. } \frac{1}{2}+1 = \frac{3}{2}\text{)}$$

Exercise 12

1. If $y = \dfrac{x^2-5x+1}{x-1}$ find $\dfrac{dy}{dx}$

2. Differentiate with respect to x, the function: $\dfrac{4x^2-x}{2x+3}$

3. Find the differential coefficient of $y = \dfrac{-7}{3x^2+1}$

4. Differentiate: $y = \dfrac{\sqrt{1+x}}{\sqrt{1-x}}$

5. Differentiate with respect to x: $y = \dfrac{(x^3-2)^2}{x^2}$

6. Find the derivative of $\dfrac{\sqrt{2+3x^2}}{x^3}$

7. If $y = \dfrac{(x^2-x-4)^{\frac{1}{3}}}{(2x+1)^2}$

8. Determine $\dfrac{d}{dx}\left(\dfrac{x-2x^3}{\sqrt{2-x}}\right)$

9. Find the differential coefficient of $y = -\dfrac{1}{1-3x^2}$

10. Differentiate: $y = \dfrac{2+x}{2-x}$

CHAPTER 13
DERIVATIVE OF PARAMETRIC EQUATIONS

If y = f(t) and $x = g(t)$ are two different functions of a common variable, t, then the two equations are called parametric equations. The variable t, is the parameter.

The derivative of a parametric equation such as the one stated above is obtained as follows:

$$\frac{dy}{dx} = \frac{\frac{dy}{dt}}{\frac{dx}{dt}}$$

Examples

1. If $y = 5 + t^2$ and $x = 3 - 2t^2$, find $\frac{dy}{dx}$

<u>Solution</u>

$$y = 5 + t^2$$

Hence, $\frac{dy}{dt} = 2t$

$$x = 3 - 2t^2$$

Hence, $\frac{dx}{dt} = -4t$

Therefore, $\frac{dy}{dx} = \frac{\frac{dy}{dt}}{\frac{dx}{dt}}$

$$= \frac{2t}{-4t}$$

$$= -\frac{1}{2} \qquad \text{(t cancels out)}$$

2. Find $\frac{dy}{dx}$ of the functions below which are expressed in the parametric form.

$$x = \frac{5}{t^2} \text{ and } y = 2t^5 - 3$$

<u>Solution</u>

$$x = \frac{5}{t^2}$$

$$x = 5t^{-2}$$

$$\frac{dx}{dt} = -10t^{-3}$$

$$y = 2t^5 - 3$$

$$\frac{dy}{dt} = 10t^4$$

Therefore, $\frac{dy}{dx} = \frac{\frac{dy}{dt}}{\frac{dx}{dt}}$

$$= \frac{10t^4}{-10t^{-3}}$$
$$= -t^{4-(-3)} \quad \text{(10 cancels out)}$$
$$= -t^{4+3}$$
$$= -t^7$$

3. If $V = \frac{4}{3}\pi r^3$ and $A = \pi r^2$, find $\frac{dA}{dV}$

Solution

$$V = \frac{4}{3}\pi r^3$$
$$\frac{dV}{dr} = 3\left(\frac{4}{3}\right)\pi r^2$$
$$= 4\pi r^2$$
$$A = \pi r^2$$
$$\frac{dA}{dr} = 2\pi r$$
$$\frac{dA}{dV} = \frac{\frac{dA}{dr}}{\frac{dV}{dr}}$$
$$= \frac{2\pi r}{4\pi r^2}$$
$$\frac{dA}{dV} = \frac{1}{2r}$$

4. Determine the derivative of the curve defined by the equations: $x = t^2 - 4t$ and $y = 2t^3 - 7t$.

Solution

$$y = 2t^3 - 7t$$

Hence, $\frac{dy}{dt} = 6t^2 - 7$

$$x = t^2 - 4t$$

Hence, $\frac{dx}{dt} = 2t - 4$

Therefore, $\frac{dy}{dx} = \frac{\frac{dy}{dt}}{\frac{dx}{dt}}$

$$= \frac{6t^2 - 7}{2t - 4}$$

5. Given that $v = u + at$ and $s = ut + \frac{1}{2}at^2$. Find $\frac{dv}{ds}$ if u and a are constants.

Solution

$$v = u + at$$

$$\frac{dv}{dt} = a$$

$$s = ut + \frac{1}{2}at^2$$

$$\frac{ds}{dt} = u + (2 \times \frac{1}{2} \times at)$$

$$= u + at$$

Hence, $\dfrac{dv}{ds} = \dfrac{\frac{dv}{dt}}{\frac{ds}{dt}}$

$$= \frac{a}{u+at}$$

6. If $y = \dfrac{t}{t-2}$ and $x = \dfrac{1}{t+1}$, find $\dfrac{dy}{dx}$

Solution

$$y = \frac{t}{t-2}$$

$$\frac{dy}{dt} = \frac{(t-2)(1) - t(1)}{(t-2)^2} \qquad \text{(Use of quotient rule where u = t and v = t – 2)}$$

$$= \frac{t-2-t}{(t-2)^2}$$

$$\frac{dy}{dt} = \frac{-2}{(t-2)^2}$$

$$x = \frac{1}{t+1}$$

$$\frac{dx}{dt} = \frac{(t+1)(0) - 1(1)}{(t+1)^2} \qquad \text{(Use of quotient rule where u = 1 and v = t + 1)}$$

$$= \frac{-1}{(t+1)^2}$$

Hence, $\dfrac{dy}{dx} = \dfrac{\frac{dy}{dt}}{\frac{dx}{dt}}$

$$= \frac{\frac{-2}{(t-2)^2}}{\frac{-1}{(t+1)^2}}$$

$$= \frac{-2}{(t-2)^2} \times \frac{(t+1)^2}{-1}$$

$$\frac{dy}{dx} = \frac{2(t+1)^2}{(t-2)^2}$$

7. The parametric equations of the motion of a stone are: $y = 12 + 3t - 2t^2$ and $x = 5t$. Find $\dfrac{dy}{dx}$.

Solution

$$y = 12 + 3t - 2t^2$$

$$\frac{dy}{dt} = 3 - 4t$$

$$x = 5t$$

$$\frac{dx}{dt} = 5$$

$$\frac{dy}{dx} = \frac{\frac{dy}{dt}}{\frac{dx}{dt}}$$

$$= \frac{3 - 4t}{5}$$

8. If the parametric equations of a parabola are y = $\frac{2mt^2}{1+t^2}$ and x = $\frac{2m}{1+t^2}$ where m is a constant, find $\frac{dy}{dx}$.

Solution

$$y = \frac{2mt^2}{1+t^2}$$

$$\frac{dy}{dt} = \frac{(1+t^2)(4mt) - 2mt^2(2t)}{(1+t^2)^2}$$ (Use of quotient rule where u = $2mt^2$ and v = 1 + t^2)

$$= \frac{4mt + 4mt^3 - 4mt^3}{(1+t^2)^2}$$

$$\frac{dy}{dt} = \frac{4mt}{(1+t^2)^2}$$

$$x = \frac{2m}{1+t^2}$$

$$\frac{dx}{dt} = \frac{(1+t^2)(0) - 2m(2t)}{(1+t^2)^2}$$ (Use of quotient rule where u = 1 and v = t + 1)

$$= \frac{-4mt}{(1+t^2)^2}$$

Hence, $\frac{dy}{dx} = \frac{\frac{dy}{dt}}{\frac{dx}{dt}}$

$$= \frac{\frac{4mt}{(1+t^2)^2}}{\frac{-4mt}{(1+t^2)^2}}$$

$$= \frac{4mt}{(1+t^2)^2} \quad x \quad \frac{(1+t^2)^2}{-4mt}$$

$$\frac{dy}{dx} = -1$$ (Same terms cancels out)

Exercise 13

1. If $y = t^3 - 5$ and $x = t^2 + 1$ find $\dfrac{dy}{dx}$

2. Find $\dfrac{dy}{dx}$ of the functions below which are expressed in the parametric form:

   $x = \dfrac{1}{4 - t^3}$ and $y = 3 + t^2$

3. If $V = \dfrac{1}{3}\pi r^2$ and $A = \pi r l + \pi r^2$, find $\dfrac{dV}{dA}$

4. Find the derivative of the curve defined by the equations: $x = 5t^2 - t + 3$ and $y = t^2 - t - 1$.

5. Given that $F = \dfrac{m(v - u)}{t}$ and $s = \dfrac{(u+v)t}{2}$. Find $\dfrac{dF}{ds}$ if u, v and m are constants.

6. If $y = \dfrac{5}{t^2 - 1}$ and $x = \dfrac{3}{t^4 + 1}$, find $\dfrac{dy}{dx}$

7. The parametric equations of the motion of a stone are: $y = t - 3t^4$ and $x = t^2 + 3$. Find $\dfrac{dy}{dx}$.

8. If the parametric equations of a parabola are $y = \dfrac{at^3}{2}$ and $x = \dfrac{at}{5}$ where a is a constant, find $\dfrac{dy}{dx}$.

9. If $V = \dfrac{1}{3}\pi r^3$ and $A = 4\pi r^2$, find $\dfrac{dA}{dV}$

10. Given that $G = 2m + s^2$ and $H = m^2 - \dfrac{3}{s}$. Find $\dfrac{dG}{dH}$ if m is a constant.

CHAPTER 14
DERIVATIVE OF IMPLICIT FUNCTIONS

In the function $y = f(x)$, y is said to be expressed explicitly in terms of x. However, in expressions such as $2xy - x^2y = 5$, the relationship between y and x is said to be implicit.

In order to differentiate implicit functions, y is differentiated just like x but with the addition of $\frac{dy}{dx}$ along with the value obtained.

Examples

1. Differentiate implicitly, the expression: $2x^2 + y^2 = 9$.

Solution

$$2x^2 + y^2 = 9$$

Follow the rule of differentiation and add $\frac{dy}{dx}$ to the value obtained whenever you differentiate y. Hence we differentiate each term in the expression above as follows:

$$\frac{d(2x^2)}{dx} + \frac{d(2y^2)}{dx} = \frac{d(9)}{dx}$$

$$4x + 2y\frac{dy}{dx} = 0 \quad \text{(The derivative of } y^2 \text{ is 2y and the addition of } \frac{dy}{dx} \text{ to 2y gives } 2y\frac{dy}{dx}\text{)}$$

We now make $\frac{dy}{dx}$ the subject of the formula as follows:

$$2y\frac{dy}{dx} = -4x$$

$$\frac{dy}{dx} = \frac{-4x}{2y} \quad \text{(When both sides are divided by 2y)}$$

$$\frac{dy}{dx} = \frac{-2x}{y}$$

2. If $x^3 + y^3 = 18xy$, find $\frac{dy}{dx}$

Solution

$$x^3 + y^3 = 18xy$$

$$3x^2 + 3y^2\frac{dy}{dx} = (18x \times 1\frac{dy}{dx}) + (y \times 18)$$

Note that $18xy$ is differentiated by using product rule where $18x$ is taken as u while y is take as v. Also, the derivative of y is what gave us $1\frac{dy}{dx}$. The above differentiation now simplifies to:

$$3x^2 + 3y^2\frac{dy}{dx} = 18x\frac{dy}{dx} + 18y$$

Collect terms in $\frac{dy}{dx}$ on one side in order to make $\frac{dy}{dx}$ the subject of the formula as follows:

$$3y^2\frac{dy}{dx} - 18x\frac{dy}{dx} = 18y - 3x^2$$

Factorizing the left hand side gives:

$$\frac{dy}{dx}(3y^2 - 18x) = 18y - 3x^2$$

Divide both sides by $3y^2 - 18x$. This gives:

$$\frac{dy}{dx} = \frac{18y - 3x^2}{3y^2 - 18x}$$

$$= \frac{3(6y - x^2)}{3(y^2 - 6x)}$$

$$\frac{dy}{dx} = \frac{6y - x^2}{y^2 - 6x} \qquad \text{(3 cancels out)}$$

3. Find $\frac{dy}{dx}$ given that $x^2y^2 - 3xy + 4xy^3 = 4$

Solution

$$x^2y^2 - 3xy + 4xy^3 = 4$$

Apply product rule to x^2y^2, $3xy$ and $4xy^3$ and differentiate appropriately as follows:

$$(x^2 \times 2y\frac{dy}{dx}) + (y^2 \times 2x) - [(3x \times 1\frac{dy}{dx}) + (y \times 3)] + (4x \times 3y^2\frac{dy}{dx}) + (y^3 \times 4) = 0$$

$$2x^2y\frac{dy}{dx} + 2xy^2 - (3x\frac{dy}{dx} + 3y) + 12xy^2\frac{dy}{dx} + 4y^3 = 0$$

$$2x^2y\frac{dy}{dx} + 2xy^2 - 3x\frac{dy}{dx} - 3y + 12xy^2\frac{dy}{dx} + 4y^3 = 0$$

$$2x^2y\frac{dy}{dx} - 3x\frac{dy}{dx} + 12xy^2\frac{dy}{dx} + 2xy^2 - 3y + 4y^3 = 0$$

$$2x^2y\frac{dy}{dx} - 3x\frac{dy}{dx} + 12xy^2\frac{dy}{dx} = 3y - 2xy^2 - 4y^3$$

Factorizing the left hand side gives:

$$\frac{dy}{dx}(2x^2y - 3x + 12xy^2) = 3y - 2xy^2 - 4y^3$$

$$\frac{dy}{dx} = \frac{3y - 2xy^2 - 4y^3}{2x^2y - 3x + 12xy^2}$$

4. Differentiate $x^4 + 6x^2y^2 - 5 = 0$ implicitly with respect to x.

Solution

$$x^4 + 6x^2y^2 - 5 = 0$$

We differentiate accordingly and apply product rule to $6x^2y^2$ as follows:

$$4x^3 + (6x^2 \times 2y\frac{dy}{dx}) + (y^2 \times 12x) - 0 = 0$$

$$4x^3 + 12x^2y\frac{dy}{dx} + 12xy^2 = 0$$

$$12x^2y\frac{dy}{dx} = -4x^3 - 12xy^2$$

$$\frac{dy}{dx} = \frac{-4x^3 - 12xy^2}{12x^2y}$$

$$= \frac{-4x(x^2 + 3y^2)}{12x^2y}$$

$$\frac{dy}{dx} = \frac{-(x^2 + 3y^2)}{3xy} \qquad \text{(4 and } x \text{ cancels out)}$$

5. Find $\dfrac{dy}{dx}$ if $\dfrac{x^2}{16} + \dfrac{y^2}{25} = 1$

Solution

$$\frac{x^2}{16} + \frac{y^2}{25} = 1$$

$$\frac{2x}{16} + \frac{2y}{25}\frac{dy}{dx} = 0$$

$$\frac{2y}{25}\frac{dy}{dx} = -\frac{2x}{16}$$

$$\frac{2y}{25}\frac{dy}{dx} = -\frac{x}{8}$$

$$\frac{dy}{dx} = \frac{-\dfrac{x}{8}}{\dfrac{2y}{25}}$$

$$= -\frac{x}{8} \times \frac{25}{2y}$$

$$\frac{dy}{dx} = -\frac{25x}{16y}$$

Exercise 14

1. Differentiate implicitly, the expression: $5x^3 + 3y = y^2$.

2. If $2x^2 + 3y^3 = 10$, find $\dfrac{dy}{dx}$

3. Find $\dfrac{dy}{dx}$ given that $xy - 3xy^2 + 4x^3 = 4$

4. Differentiate $2x^2 + xy^2 - 5 = x^3$ implicitly with respect to x.

5. Find $\dfrac{dy}{dx}$ if $\dfrac{x^3}{3} - \dfrac{y^2}{2} = 0$

6. Differentiate implicitly, the expression: $2x^2y + 5y = 1$.

7. If $2x^2y + 4y^3 = 2y$, find $\dfrac{dy}{dx}$

8. Find $\dfrac{dy}{dx}$ given that $5y^2 - 4xy = y$

9. Differentiate $x + y^2 - x^2y^2 = 7$ implicitly with respect to x.

10. Find $\dfrac{dy}{dx}$ if $x^5 - \dfrac{3}{y} = 2$

CHAPTER 15
DERIVATIVE OF TRIGONOMETRIC FUNCTIONS

The derivatives of trigonometric functions are as given below:

If y = sin$x$, then $\dfrac{dy}{dx}$ = cos$x$

If y = cos$x$, then $\dfrac{dy}{dx}$ = $-$sin$x$

If y = tan$x$, then $\dfrac{dy}{dx}$ = sec^2x

If y = cot$x$, then $\dfrac{dy}{dx}$ = $-$cosec^2x

If y = sec$x$, then $\dfrac{dy}{dx}$ = sec$x$tan$x$

If y = cosec$x$, then $\dfrac{dy}{dx}$ = $-$cot$x$cosec$x$

Example

1. Find the derivative of cos5x

Solution

    Let y = cos5x

We have to use chain rule from composite function due to 5x which is a function of x.

Hence, let u = 5x

Therefore, y = cosu   (By replacing 5x with u)

    $\dfrac{du}{dx}$ = 5

    $\dfrac{dy}{du}$ = $-$sinu   (Recall that the derivative of cos$x$ is $-$sin$x$)

Therefore, $\dfrac{dy}{dx}$ = $\dfrac{dy}{du}$ x $\dfrac{du}{dx}$ (Chain rule)

    = $-$sinu x 5

    = $-$5sinu

    $\dfrac{dy}{dx}$ = $-$5sin5x (Since u = 5x)

2. If y = sin$\dfrac{1}{2}$$x$, find $\dfrac{dy}{dx}$

Solution

    y = sin$\dfrac{1}{2}$$x$

Let u = $\dfrac{1}{2}$$x$

Therefore, y = sinu   (By replacing $\dfrac{1}{2}$$x$ with u)

$$\frac{du}{dx} = \frac{1}{2}$$

$$\frac{dy}{du} = \cos u \quad \text{(Recall that the derivative of } \sin x \text{ is } \cos x)$$

Therefore, $\dfrac{dy}{dx} = \dfrac{dy}{du} \times \dfrac{du}{dx}$ (Chain rule)

$$= \cos u \times \frac{1}{2}$$

$$= \frac{1}{2}\cos u$$

$$\frac{dy}{dx} = \frac{1}{2}\cos\frac{1}{2}x \quad \text{(Since } u = \frac{1}{2}x)$$

3. Find the derivative of $5\cos 3x$

<u>Solution</u>

$$y = 5\cos 3x$$

Let $u = 3x$

Therefore, $y = 5\cos u$

$$\frac{du}{dx} = 3$$

$$\frac{dy}{du} = 5(-\sin u) \quad \text{(Note that the constant term i.e. 5 should be used to multiply the derivative)}$$

$$= -5\sin u$$

Therefore, $\dfrac{dy}{dx} = \dfrac{dy}{du} \times \dfrac{du}{dx}$

$$= -5\sin u \times 3$$

$$= -15\sin u$$

$$\frac{dy}{dx} = -15\sin 3x \quad \text{(Since } u = 3x)$$

4. Find the derivative of $\sin^2 x$

<u>Solution</u>

$$y = \sin^2 x$$

Note that $\sin^2 x = \sin x \times \sin x$

Hence, let $u = \sin x$

Therefore, $y = u^2$ (i.e. $u \times u$ from $\sin x \times \sin x$)

$$\frac{du}{dx} = \cos x$$

$$\frac{dy}{du} = 2u$$

Therefore, $\dfrac{dy}{dx} = \dfrac{dy}{du} \times \dfrac{du}{dx}$

$$= 2u \times \cos x$$

$$= 2u\cos x$$

$$\frac{dy}{dx} = 2\sin x\cos x \qquad \text{(Since } u = \sin x\text{)}$$

5. Differentiate with respect to x: $y = \tan 2x$

<u>Solution</u>

$$y = \tan 2x$$

$$\frac{dy}{dx} = \sec^2 2x \times \frac{d(2x)}{dx} \qquad \text{(Note that the derivative of } \tan x \text{ is } \sec^2 x\text{)}$$

$$= \sec^2 2x \times 2$$

$$= 2\sec^2 2x$$

6. If $y = \cos^3 6x^2$, differentiate y with respect to x.

<u>Solution</u>

$$y = \cos^3 6x^2$$

Let us solve this problem without the use of v as follows:

Let $u = \cos 6x^2$

Hence, $y = u^3$

$$\frac{du}{dx} = 12x(-\sin 6x^2) \qquad \text{(Note that } 12x \text{ is from the derivative of } 6x^2\text{)}$$

$$= -12x\sin 6x^2$$

$$\frac{dy}{du} = 3u^2$$

Therefore, $\dfrac{dy}{dx} = \dfrac{dy}{du} \times \dfrac{du}{dx}$

$$= 3u^2 \times (-12x\sin 6x^2)$$

$$= -36xu^2\sin 6x^2$$

$$= -36x(\cos 6x^2)^2\sin 6x^2 \qquad \text{(u has been replaced with } \cos 6x^2\text{)}$$

$$\frac{dy}{dx} = -36x\cos^2 6x^2\sin 6x^2$$

7. Find the derivative of $y = \sec 3x$

<u>Solution</u>

$$y = \sec 3x$$

$$\frac{dy}{dx} = \sec 3x\tan 3x \times \frac{d(3x)}{dx} \qquad \text{(Note that the derivative of } \sec x \text{ is } \sec x\tan x\text{)}$$

$$= \sec 3x\tan 3x \times 3$$

$$\frac{dy}{dx} = 3\sec 3x\tan 3x$$

8. Find the derivative of $\csc 4x^3$

Solution

$$y = \text{cosec}4x^3$$

$$\frac{dy}{dx} = -\text{cosec}4x^3\text{cot}4x^3 \times \frac{d(4x^3)}{dx}$$ (Note that the derivative of cosec is −cosec cot)

$$= -\text{cosec}4x^3\text{cot}4x^3 \times 12x^2$$

$$= -12x^2\text{cosec}4x^3\text{cot}4x^3$$

9. Find the derivative of $\text{cot}^2 2x^4$

Solution

$$y = \text{cot}^2 2x^4$$

This can also be written as: $y = \text{cot}2x^4 \times \text{cot}2x^4$

Let $u = 2x^4$ (Take the function of x)

Also, let $v = \text{cot } u$ (Take a function of u without taking the exponent)

Hence, $y = v^2$ (Since $v^2 = (\text{cot } u)^2 = (\text{cot}2x^4)^2 = \text{cot}^2 2x^4$. Hence, $y = v^2$)

$$\frac{du}{dx} = 8x^3$$

$$\frac{dv}{du} = -\text{cosec}^2 u$$

$$\frac{dy}{dv} = 2v$$

Therefore, $\dfrac{dy}{dx} = \dfrac{dy}{dv} \times \dfrac{dv}{du} \times \dfrac{du}{dx}$

$$= 2v \times -\text{cosec}^2 u \times 8x^3$$

$$= -16x^3 v\,\text{cosec}^2 u$$

$$= -16x^3 \text{cot } u\,\text{cosec}^2 u$$ (Since $v = \text{cot } u$)

Substituting in the original value of u gives:

$$\frac{dy}{dx} = -16x^3\text{cot }2x^4\text{cosec}^2 2x^4$$

10. Find the derivative of $5x\text{sin}2x$

Solution

$$y = 5x\text{sin}2x$$

We are going to apply the product rule of differentiation.

Let $u = 5x$

and $v = \text{sin}2x$

$$\frac{du}{dx} = 5$$

$$\frac{dv}{dx} = 2\text{cos}2x$$ (Note that the differentiation of $2x$ gives 2)

Hence, $\dfrac{dy}{dx} = u\dfrac{dv}{dx} + v\dfrac{du}{dx}$

$$= (5x \times 2\text{cos}2x) + (\text{sin}2x \times 5)$$

$$= 10x\cos 2x + 5\sin 2x$$

$$\frac{dy}{dx} = 5(2x\cos 2x + \sin 2x)$$

11. Find $\frac{dy}{dx}$ if $y = \frac{1}{x}\sec x$

<u>Solution</u>

$$y = \frac{1}{x}\sec x$$

We apply product rule as follows:

$$u = \frac{1}{x}$$

$$= x^{-1}$$

$$v = \sec 3x$$

$$\frac{du}{dx} = -1x^{-2}$$

$$= \frac{-1}{x^2}$$

$$\frac{dv}{dx} = \sec 3x\tan 3x \ \times \ 3$$

$$= 3\sec 3x\tan 3x$$

Note that the derivative of $\sec x$ is $\sec x\tan x$ and 3 is from the derivative of $3x$

Hence, $\frac{dy}{dx} = u\frac{dv}{dx} + v\frac{du}{dx}$

$$= \frac{1}{x}(3\sec 3x\tan 3x) + \sec 3x\left(\frac{-1}{x^2}\right)$$

$$\frac{dy}{dx} = \frac{3}{x}\sec 3x\tan 3x - \frac{1}{x^2}\sec 3x$$

12. Find the derivative of $3\csc x^6$

<u>Solution</u>

$$y = 3\csc x^6$$

Let $u = x^6$

Hence, $y = 3\csc u$

$$\frac{du}{dx} = 6x^5$$

$$\frac{dy}{du} = -3\cot u \csc u \quad \text{(The constant term i.e. 5 should be used to multiply the derivative)}$$

Therefore, $\frac{dy}{dx} = \frac{dy}{du} \ \times \ \frac{du}{dx}$

$$= -3\cot u \csc u \ \times \ 6x^5$$

$$= -18x^5\cot u \csc u$$

$$\frac{dy}{dx} = -18x^5\cot x^6 \csc x^6 \quad \text{(Since } u = x^6\text{)}$$

13. If $y = \dfrac{1 + \cos 2x}{\sin 2x}$ find $\dfrac{dy}{dx}$.

Solution

$$y = \frac{1 + \cos 2x}{\sin 2x}$$

We have to apply quotient rule on this as follows:

$u = 1 + \cos2x$

$v = \sin2x$

$$\frac{du}{dx} = -2\sin2x$$

$$\frac{dv}{dx} = 2\cos2x$$

Hence, $\dfrac{dy}{dx} = \dfrac{v\frac{du}{dx} - u\frac{dv}{dx}}{v^2}$ (Quotient rule)

$$= \frac{\sin 2x(-2\sin 2x) - (1 + \cos 2x)(2\cos 2x)}{(\sin 2x)^2}$$

$$= \frac{-2\sin^2 2x - (2\cos 2x + 2\cos^2 2x)}{\sin^2 2x}$$ (Note that $(\sin2x)^2 = \sin^2 2x$)

$$= \frac{-2\sin^2 2x - 2\cos 2x - 2\cos^2 2x}{\sin^2 2x}$$

$$= \frac{-2\sin^2 2x - 2\cos^2 2x - 2\cos 2x}{\sin^2 2x}$$

$$= \frac{-2(\sin^2 2x + \cos^2 2x) - 2\cos 2x}{\sin^2 2x}$$

$$= \frac{-2(1) - 2\cos 2x}{\sin^2 2x}$$ (Note that $\sin^2 x + \cos^2 x = 1$, hence $\sin^2 2x + \cos^2 2x = 1$)

$$\frac{dy}{dx} = \frac{-2(1 + \cos 2x)}{\sin^2 2x}$$

14. Differentiate $y = \dfrac{\sin^2 x}{x}$

Solution

$$y = \frac{\sin^2 x}{x}$$

This is also quotient rule.

Therefore, $u = \sin^2 x$

$v = x$

Let us follow a direct and systematic way of differentiating trigonometric functions.

Hence, in order to differentiate $\sin 2x$:

First, differentiate the exponent of $\sin^2$ without changing the trigonometric term. This gives:

$2\sin^{2-1}$

$= 2\sin$

201

Then add the term in x. This gives:

$2\sin x$

The next step is to differentiate sin which gives cos. Also add the term in x to obtain $\cos x$.

Finally, differentiate the term in x. Hence, differentiating x gives 1.

Now multiply the three terms obtained in the three steps above. This gives:

$2\sin x \ \times \ \cos x \ \times \ 1$

$= 2\sin x\cos x$

Hence, $\dfrac{du}{dx} = 2\sin x\cos x$ \qquad (As obtained above)

Since, $v = x$

Then, $\dfrac{dv}{dx} = 1$

Hence, $\dfrac{dy}{dx} = \dfrac{v\dfrac{du}{dx} - u\dfrac{dv}{dx}}{v^2}$

$= \dfrac{x(2\sin x\cos x) - \sin^2 x(1)}{x^2}$

$= \dfrac{2x\sin x\cos x - \sin^2 x}{x^2}$

$= \dfrac{\sin x(2x\cos x - \sin x)}{x^2}$

15. Differentiate with respect to x: $\dfrac{\sec 2x}{x^3 + 1}$

<u>Solution</u>

$y = \dfrac{\sec 2x}{x^3 + 1}$

$u = \sec 2x$

$v = x^3 + 1$

Hence, $\dfrac{du}{dx} = 2\sec 2x\tan 2x$

Then, $\dfrac{dv}{dx} = 3x^2$

Hence, $\dfrac{dy}{dx} = \dfrac{v\dfrac{du}{dx} - u\dfrac{dv}{dx}}{v^2}$

$= \dfrac{(x^3 + 1)(2\sec 2x\tan 2x) - \sec 2x(3x^2)}{(x^3 + 1)^2}$

$= \dfrac{(x^3 + 1)(2\sec 2x\tan 2x) - 3x^2\sec 2x}{(x^3 + 1)^2}$

16. Find the derivative of $\dfrac{\cos \sqrt{x}}{1 + x}$

Solution

$$y = \frac{\cos\sqrt{x}}{1+x}$$

$$u = \cos\sqrt{x}$$

$$= \cos x^{\frac{1}{2}}$$

$$v = 1 + x$$

Hence, $\dfrac{du}{dx} = \dfrac{d(x^{\frac{1}{2}})}{dx} \times \dfrac{d(\cos)}{dx}$

Note that $\dfrac{d(\cos)}{dx}$ means the derivative of cos which gives $-\sin$, and then $x^{\frac{1}{2}}$ is added to it to give $-\sin x^{\frac{1}{2}}$. Hence we continue as follows:

$$\frac{du}{dx} = \frac{1}{2}x^{-\frac{1}{2}} \times -\sin x^{\frac{1}{2}}$$

$$= \frac{1}{2x^{\frac{1}{2}}} \times -\sin x^{\frac{1}{2}}$$

$$\frac{du}{dx} = \frac{-\sin x^{\frac{1}{2}}}{2x^{\frac{1}{2}}}$$

Also, $\dfrac{dv}{dx} = 1$

Hence, $\dfrac{dy}{dx} = \dfrac{v\frac{du}{dx} - u\frac{dv}{dx}}{v^2}$

$$= \frac{(1+x)\left(\dfrac{-\sin x^{\frac{1}{2}}}{2x^{\frac{1}{2}}}\right) - \cos x^{\frac{1}{2}}(1)}{(1+x)^2}$$

$$= \frac{-\sin\sqrt{x}}{2\sqrt{x}(1+x)} - \frac{\cos\sqrt{x}}{(1+x)^2} \qquad \text{(When the fractions are separated)}$$

$$\frac{dy}{dx} = \frac{-(1+x)\sin\sqrt{x} - 2\sqrt{x}\cos\sqrt{x}}{2\sqrt{x}(1+x)^2} \qquad \text{(When the fractions are combined)}$$

17. Find $\dfrac{dy}{dx}$ if $y = \dfrac{1-x^2}{1+\cos x}$

Solution

$$y = \frac{1-x^2}{1+\cos x}$$

$$u = 1 - x^2$$

$$v = 1 + \cos x$$

$$\frac{du}{dx} = -2x$$

$$\frac{dv}{dx} = -\sin x$$

Hence, $\dfrac{dy}{dx} = \dfrac{v\frac{du}{dx} - u\frac{dv}{dx}}{v^2}$

$= \dfrac{(1+\cos x)(-2x) - (1-x^2)(-\sin x)}{(1+\cos x)^2}$

$\dfrac{dy}{dx} = \dfrac{-2x(1+\cos x) + (1-x^2)\sin x}{(1+\cos x)^2}$

Note that the negative sign from −sin$x$ changed the negative sign at the middle to a positive sign since negative sign multiplied by negative sign gives a positive sign.

18. Find the derivative of $8x\sin x^2$

Solution

$y = 8x\sin x^2$

We apply product rule as follows:

$u = 8x$

$\dfrac{du}{dx} = 8$

$v = \sin x^2$

$\dfrac{dv}{dx} = \dfrac{d(\sin)}{dx} \times \dfrac{d(x^2)}{dx}$

$\quad = \cos x^2 \times 2x$

$\quad = 2x\cos x^2$

Note that $\dfrac{d(\sin)}{dx}$ gives cos which result to cosx^2 when x^2 from the question is added

Hence, $\dfrac{dy}{dx} = u\dfrac{dv}{dx} + v\dfrac{du}{dx}$

$\quad = 8x(2x\cos x^2) + \sin x^2(8)$

$\quad = 16x^2\cos x^2 + 8\sin x^2$

$\dfrac{dy}{dx} = 8(2x^2\cos x^2 + \sin x^2)$

19. Differentiate with respect to x: $\sec 2x\sin^3 2x$

Solution

$y = \sec 2x\sin^3 2x$

We apply product rule as follows:

$u = \sec 2x$

$\dfrac{du}{dx} = \dfrac{d(\sec)}{dx} \times \dfrac{d(2x)}{dx}$

$\quad = \sec 2x\tan 2x \times 2$

$\dfrac{du}{dx} = 2\sec 2x\tan 2x$

204

Note that $\dfrac{d(\sec)}{dx}$ gives sectan, but remember to add $2x$ after sec and tan respectively to obtain

sec2xtan2x

$\qquad$ v = sin^{3}2x

In order to directly differentiate a trigonometric term with exponent like this (i.e. sin^3) we have three solutions to multiply as follows:

First solution: consider only the exponent and differentiate sin^3. This gives:

$\qquad \dfrac{d(\sin^3)}{dx} = 3\sin^2$

We now add $2x$ from the question to obtain 3sin^{2}2x

Second solution: $\dfrac{d(\sin)}{dx} = \cos$ which gives cos2x

Third solution: $\dfrac{d(2x)}{dx} = 2$

Multiply theses three solutions to give the derivative of v = sin^{3}2x as follows:

$\qquad \dfrac{dv}{dx} = 3\sin^2 2x \times \cos 2x \times 2$

$\qquad\qquad = 6\sin^2 2x\cos 2x$

Now, $\dfrac{dy}{dx} = u\dfrac{dv}{dx} + v\dfrac{du}{dx}$

$\qquad = \sec 2x(6\sin^2 2x\cos 2x) + \sin^3 2x(2\sec 2x\tan 2x)$

$\qquad = \dfrac{1}{\cos 2x}(6\sin^2 2x\cos 2x) + \sin^3 2x(2\dfrac{1}{\cos 2x}\dfrac{\sin 2x}{\cos 2x})$ (Note that $\sec 2x = \dfrac{1}{\cos 2x}$ and $\tan 2x = \dfrac{\sin 2x}{\cos 2x}$)

$\qquad = 6\sin^2 2x + 2\sin^2 2x(\dfrac{\sin 2x}{\cos 2x}\dfrac{\sin 2x}{\cos 2x})$

Note that one sin2x has been taken out of sin^{3}2x and placed inside the bracket.

$\qquad = 6\sin^2 2x + 2\sin^2 2x(\tan 2x\tan x)$ (Since $\dfrac{\sin 2x}{\cos 2x} = \tan 2x$)

$\qquad = 6\sin^2 2x + 2\sin^2 2x(\tan^2 2x)$

$\dfrac{dy}{dx} = 2\sin^2 2x(3 + \tan^2 2x)$ (After factorization)

20. Differentiate with respect to x: $\dfrac{\sec x - \tan x}{\sec x + \tan x}$

<u>Solution</u>

$\qquad y = \dfrac{\sec x - \tan x}{\sec x + \tan x}$

$\qquad u = \sec x - \tan x$

$\qquad \dfrac{du}{dx} = \sec x\tan x - \sec^2 x$

$\qquad v = \sec x + \tan x$

$\qquad \dfrac{dv}{dx} = \sec x\tan x + \sec^2 x$

$$\frac{dy}{dx} = \frac{v\frac{du}{dx} - u\frac{dv}{dx}}{v^2}$$

$$= \frac{(\sec x + \tan x)(\sec x\tan x - \sec 2x) - (\sec x - \tan x)(\sec x\tan x + \sec 2x)}{(\sec x + \tan x)^2}$$

Expanding bracket in the numerator gives:

$$\frac{dy}{dx} = \frac{\sec^2 x\tan x - \sec^3 x + \sec x\tan^2 x - \sec^2 x\tan x - [\sec^2 x\tan x + \sec^3 x - \sec x\tan^2 x - \sec^2 x\tan x]}{(\sec x + \tan x)^2}$$

$$= \frac{\sec^2 x\tan x - \sec^3 x + \sec x\tan^2 x - \sec^2 x\tan x - \sec^2 x\tan x - \sec^3 x + \sec x\tan^2 x + \sec^2 x\tan x]}{(\sec x + \tan x)^2}$$

$$= \frac{2\sec x\tan^2 x - 2\sec^3 x}{(\sec x + \tan x)^2} \qquad \text{(Note that all the sec}^2x\text{tan}x \text{ have cancelled out)}$$

$$= \frac{2\sec x(\tan^2 x - \sec^2 x)}{(\sec x + \tan x)^2}$$

$$= \frac{2\sec x(-1)}{(\sec x + \tan x)^2} \qquad \text{(Note that tan}^2x - \text{sec}^2x = -1\text{)}$$

$$= \frac{-2\sec x}{(\sec x + \tan x)^2}$$

$$\frac{dy}{dx} = -\frac{2\sec x}{(\sec x + \tan x)^2}$$

21. Find the derivative of $\sqrt{\dfrac{\cos 2x}{1 + \sin 2x}}$

Solution

$$y = \sqrt{\frac{\cos 2x}{1 + \sin 2x}}$$

This can also be written as:

$$y = \frac{(\cos 2x)^{\frac{1}{2}}}{(1 + \sin 2x)^{\frac{1}{2}}}$$

$$u = (\cos 2x)^{\frac{1}{2}}$$

Using the chain rule, we obtain $\dfrac{du}{dx}$ as follows:

$$\frac{du}{dx} = \frac{1}{2}(\cos 2x)^{\frac{1}{2}-1} \times \frac{d(\cos 2x)}{dx}$$

$$= \frac{1}{2}(\cos 2x)^{-\frac{1}{2}} \times -2\sin 2x$$

$$= \frac{-2\sin 2x}{2(\cos 2x)^{\frac{1}{2}}}$$

$$= \frac{-\sin 2x}{(\cos 2x)^{\frac{1}{2}}}$$

$$v = (1 + \sin 2x)^{\frac{1}{2}}$$

206

$$\frac{dv}{dx} = \frac{1}{2}(1 + \sin 2x)^{\frac{1}{2}-1} \times \frac{d(1+\sin 2x)}{dx}$$

$$= \frac{1}{2}(1 + \sin 2x)^{-\frac{1}{2}} \times 2\cos 2x$$

$$= \frac{2\cos 2x}{2(1 + \sin 2x)^{\frac{1}{2}}}$$

$$= \frac{\cos 2x}{(1 + \sin 2x)^{\frac{1}{2}}}$$

$$\frac{dy}{dx} = \frac{v\frac{du}{dx} - u\frac{dv}{dx}}{v^2}$$

$$= \frac{(1 + \sin 2x)^{\frac{1}{2}}\left(\frac{-\sin 2x}{(\cos 2x)^{\frac{1}{2}}}\right) - (\cos 2x)^{\frac{1}{2}}\left(\frac{\cos 2x}{(1 + \sin 2x)^{\frac{1}{2}}}\right)}{[(1 + \sin 2x)^{\frac{1}{2}}]^2}$$

$$= \frac{\left(\frac{-\sin 2x(1 + \sin 2x)^{\frac{1}{2}}}{(\cos 2x)^{\frac{1}{2}}}\right) - \left(\frac{\cos 2x(\cos 2x)^{\frac{1}{2}}}{(1 + \sin 2x)^{\frac{1}{2}}}\right)}{1 + \sin 2x}$$

$$= \frac{\left(\frac{-\sin 2x(1 + \sin 2x) - \cos 2x(\cos 2x)}{(\cos 2x)^{\frac{1}{2}}(1 + \sin 2x)^{\frac{1}{2}}}\right)}{1 + \sin 2x}$$

Note that $(\cos 2x)^{\frac{1}{2}} \times (\cos 2x)^{\frac{1}{2}} = \cos 2x$ (Since the exponents are added). Similarly:

$(1 + \sin 2x)^{\frac{1}{2}} \times (1 + \sin 2x)^{\frac{1}{2}} = 1 + \sin 2x$. Simplifying the above expression further, gives:

$$\frac{dy}{dx} = \frac{-\sin 2x(1 + \sin 2x) - \cos^2 2x}{(\cos 2x)^{\frac{1}{2}}(1 + \sin 2x)^{\frac{1}{2}}(1 + \sin 2x)}$$

$$= \frac{-\sin 2x - \sin^2 2x - \cos^2 2x}{(\cos 2x)^{\frac{1}{2}}(1 + \sin 2x)^{\frac{1}{2}}(1 + \sin 2x)}$$

$$= \frac{-\sin 2x - (\sin^2 2x + \cos^2 2x)}{(\cos 2x)^{\frac{1}{2}}(1 + \sin 2x)^{\frac{1}{2}}(1 + \sin 2x)}$$

$$= \frac{-\sin 2x - 1}{(\cos 2x)^{\frac{1}{2}}(1 + \sin 2x)^{\frac{1}{2}}(1 + \sin 2x)} \quad \text{(Note that } \sin^2 2x + \cos^2 2x = 1\text{)}$$

$$= \frac{-(1 + \sin 2x)}{(\cos 2x)^{\frac{1}{2}}(1 + \sin 2x)^{\frac{1}{2}}(1 + \sin 2x)}$$

$$= \frac{-1}{(\cos 2x)^{\frac{1}{2}}(1 + \sin 2x)^{\frac{1}{2}}} \quad \text{(Note that } 1 + \sin 2x \text{ cancels out)}$$

207

22. Differentiate with respect to x: $\sin x - 2x\cos x$

Solution

$\qquad y = \sin x - 2x\cos x$

Treat $2x\cos x$ using product rule.

$$\frac{dy}{dx} = \frac{d(\sin x)}{dx} - \left(2x\,\frac{d(\cos x)}{dx} + \cos x\,\frac{d(2x)}{dx}\right)$$

$$= \cos x - [2x(-\sin x) + \cos x(2)]$$

$$= \cos x + 2x\sin x - 2\cos x$$

$$= 2x\sin x - \cos x$$

23. Find the derivative of $\cos^5 3x^4$

Solution

$\qquad y = \cos^5 3x^4$

A direct way of differentiating this problem is applied as follows:

$$\frac{dy}{dx} = \frac{d(\cos^5)}{dx} \times \frac{d(\cos)}{dx} \times \frac{d(3x^4)}{dx}$$

$$= 5\cos^4 3x^4 \times (-\sin 3x^4) \times 12x^3$$

Note that the derivative of $\cos^5$ gives $5\cos^4$ (do not change the trigonometric term, i.e. cos) and then the addition of $3x^4$ from the question gives $5\cos^4 3x^4$.

Similarly the derivative of cos gives $-\sin$ and the addition of $3x^4$ from the question gives $-\sin 3x^4$. Hence, multiplying the terms above gives:

$$\frac{dy}{dx} = -60x^3\cos^4 3x^4\sin 3x^4$$

24. Differentiate with respect to x: $\sin^8 15x^6$

Solution

$\qquad y = \sin^8 15x^6$

We can also differentiate this problem directly as follows:

$$\frac{dy}{dx} = \frac{d(\sin^8)}{dx} \times \frac{d(\sin)}{dx} \times \frac{d(15x^6)}{dx}$$

$$= 8\sin^7 15x^6 \times \cos 15x^6 \times 90x^5$$

Note that the derivative of $\sin^8$ gives $8\sin^7$ (in this case do not change the trigonometric term, i.e. sin) and then the addition of $15x^6$ from the question gives $8\sin^7 15x^6$.

Similarly the derivative of sin gives cos and the addition of $15x^6$ from the question gives $\cos 15x^6$. Hence, we continue as follows:

$$\frac{dy}{dx} = 8\sin^7 15x^6 \times \cos 15x^6 \times 90x^5$$

$$= 720x^5\sin^7 15x^6\cos 15x^6$$

Exercise 15

1. Find the derivative of $\tan 2x$

2. If $y = \cos \frac{1}{5} x$, find $\frac{dy}{dx}$

3. Find the derivative of $10\cos 5x$

4. Find the derivative of $\sin^3 x$

5. Differentiate with respect to x: $y = \tan^2 x$

6. If $y = \sin^4 3x^5$, differentiate y with respect to x.

7. Find the derivative of $y = \csc 6x^2$

8. Find the derivative of $\sec 2x^5$

9. Find the derivative of $\tan 3x^2$

10. Find the derivative of $x^2 \cos 3x$

11. Find $\frac{dy}{dx}$ if $y = \frac{3}{x^3} \sin 2x$

12. Find the derivative of $12\sec x^4$

13. If $y = \dfrac{2 - \sin 5x}{\tan x}$ find $\frac{dy}{dx}$.

14. Differentiate $y = \dfrac{\sin^3 x}{2x}$

15. Differentiate with respect to x: $\dfrac{\cot 2x}{x + 3}$

16. Find the derivative of $\dfrac{\sin \sqrt[3]{3}}{5x}$

17. Find $\frac{dy}{dx}$ if $y = \dfrac{x^2 - 3}{\sec 2x}$

18. Find the derivative of $3x^2 \cos 3x^2$

19. Differentiate with respect to x: $\cot x \cos^2 x$

20. Differentiate with respect to x: $\dfrac{\sin x - \cos x}{\sin x + \cos x}$

21. Find the derivative of $\dfrac{\sec 4x}{\cos 4x - 2}$

22. Differentiate with respect to x: $\cos 3x - x^2 \sec x$

23. Find the derivative of $\sin^9 x^3$

24. Differentiate with respect to x: $\sin 6x^{\frac{1}{2}}$

25. Find $\frac{dy}{dx}$ if $y = \frac{1}{x^2} \sin x^3$

26. Find the derivative of $\cos^3 2x^5$

27. If $y = \dfrac{\cos 10x}{\sin 2x}$ find $\frac{dy}{dx}$.

28. Differentiate $y = \dfrac{3\sin x^4}{2x}$

29. Differentiate with respect to x: $\dfrac{\tan 5x}{2x - 1}$

30. Find the derivative of $\dfrac{\tan^2 x}{2x}$

CHAPTER 16
DERIVATIVE OF INVERSE FUNCTIONS

If the derivative of a function is given by $\dfrac{dy}{dx}$, then the derivative of the inverse function is given

by: $\dfrac{1}{\frac{dy}{dx}} = \dfrac{dx}{dy}$

Or, $\dfrac{dy}{dx} = \dfrac{1}{\frac{dx}{dy}}$

Examples

1. Find $\dfrac{dx}{dy}$ if y = $\sqrt[3]{x}$

<u>Solution</u>

Method 1

$\quad$ y = $\sqrt[3]{x}$

Or $\quad$ y = $x^{\frac{1}{3}}$

Let us make x the subject of the formula. The inverse of $\dfrac{1}{3}$ is 3. Hence raise both sides to the

exponent 3 as follows:

$\quad y^3 = (x^{\frac{1}{3}})^3$

$\quad y^3 = x^1 \quad$ (Note that 1 was obtained from $\dfrac{1}{3}$ x 3)

Hence, $x = y^3$

Therefore, $\dfrac{dx}{dy} = 3y^2$

Method 2

$\quad$ y = $\sqrt[3]{x}$

Or $\quad$ y = $x^{\frac{1}{3}}$

$\dfrac{dy}{dx} = \dfrac{1}{3}x^{\frac{1}{3}-1}$

$\quad\quad = \dfrac{1}{3}x^{-\frac{2}{3}}$

$\dfrac{dy}{dx} = \dfrac{1}{3x^{\frac{2}{3}}}$

Hence, $\dfrac{dx}{dy} = \dfrac{1}{\frac{dy}{dx}}$

$\quad = \dfrac{3x^{\frac{2}{3}}}{1} \quad$ (This means the inverse of $\dfrac{1}{3x^{\frac{2}{3}}}$)

$$= 3x^{\frac{2}{3}}$$

$$= 3(\sqrt[3]{x})^2 \qquad \text{[Recall from indices that } x^{\frac{a}{b}} = (\sqrt[b]{x})^a \text{]}$$

$$\frac{dx}{dy} = 3y^2 \qquad \text{(Since } y = \sqrt[3]{x})$$

2. If $y = \sqrt[5]{2x - 3}$ find $\dfrac{dx}{dy}$

Solution

$$y = \sqrt[5]{2x - 3}$$

Or $y = (2x - 3)^{\frac{1}{5}}$

Let us make x the subject of the formula. The inverse of $\dfrac{1}{5}$ is 5. Hence raise both sides to the exponent 5 as follows:

$$y^5 = [(2x - 3)^{\frac{1}{5}}]^5$$

$$y^5 = 2x - 3 \qquad \text{(Note that } \dfrac{1}{5} \text{ x 5 = 1, which cancels the fractional exponent)}$$

Hence, $2x = y^5 + 3$

$$x = \frac{y^5 + 3}{2}$$

$$x = \frac{y^5}{2} + \frac{3}{2} \qquad \text{(When we separate into fractions by dividing each part by the denominator)}$$

Therefore, $\dfrac{dx}{dy} = \dfrac{5y^4}{2}$

3. If $y = x^3 - 5$. Find $\dfrac{dx}{dy}$

Solution

$$y = x^3 - 5$$
$$y + 5 = x^3$$
$$x^3 = y + 5$$
$$x = (y + 5)^{\frac{1}{3}} \qquad \text{(By raising both sides to an exponent of the inverse of 3 which is } \dfrac{1}{3})$$

$$\frac{dx}{dy} = \frac{1}{3}(y + 5)^{\frac{1}{3} - 1} \text{ x } 1$$

Note that the 1 was obtained from the derivative of y + 5 since chain rule was used.

$$\frac{dx}{dy} = \frac{1}{3}(y + 5)^{-\frac{2}{3}}$$

$$\frac{dx}{dy} = \frac{1}{3(y+5)^{\frac{2}{3}}}$$

Recall from indices that $x^{\frac{a}{b}} = (\sqrt[b]{x})^a$. Applying this rule gives:

$$\frac{dx}{dy} = \frac{1}{3[\sqrt[3]{(y+5)^2}]}$$

4. If $y = \frac{1}{2}x^4 + 3$, find $\frac{dx}{dy}$

Solution

$$y = \frac{1}{2}x^4 + 3$$

$$y - 3 = \frac{1}{2}x^4$$

$$2(y - 3) = x^4$$

$$x^4 = 2y - 6$$

$x = (2y - 6)^{\frac{1}{4}}$ (This is obtained by raising both sided to an exponent of the inverse of 4, i.e. $\frac{1}{4}$)

Hence we use chain rule to determine $\frac{dx}{dy}$ as follows:

$$\frac{dx}{dy} = \frac{1}{4}(2y - 6)^{\frac{1}{4} - 1} \times 2 \qquad \text{(Note that 2 is from the derivative of } 2y - 6)$$

$$\frac{dx}{dy} = \frac{1}{2}(2y - 6)^{-\frac{3}{4}}$$

$$\frac{dx}{dy} = \frac{1}{2(2y - 6)^{\frac{3}{4}}}$$

Hence, $\frac{dx}{dy} = \frac{1}{2[\sqrt[4]{(2y-6)^3}]}$ [This is obtained from the law of indices given by: $x^{\frac{a}{b}} = (\sqrt[b]{x})^a$]

5. If $y = \frac{x + 2}{x}$, find $\frac{dx}{dy}$.

Solution

$$y = \frac{x + 2}{x}$$

$xy = x + 2$ (When we cross multiply)

$$xy - x = 2$$

Factorizing the left hand side gives:

$$x(y - 1) = 2$$

$$x = \frac{2}{y - 1}$$

$x = 2(y - 1)^{-1}$ (Take note of the use of negative exponent when the denominator goes up)

Hence we use chain rule to determine $\frac{dx}{dy}$ as follows:

$$\frac{dx}{dy} = -1 \times 2(y - 1)^{-1-1} \times 1 \qquad \text{(Note that 1 is from the derivative of } y - 1)$$

213

$$\frac{dx}{dy} = -2(y-1)^{-2}$$

$$\frac{dx}{dy} = \frac{-2}{(y-1)^2}$$

6. Find $\frac{dx}{dy}$ if $y = \frac{1}{x+2}$

Solution

$$y = \frac{1}{x+2}$$

$$x + 2 = \frac{1}{y}$$

$$x = \frac{1}{y} - 2$$

$$x = y^{-1} - 2$$

$$\frac{dx}{dy} = -1 \times y^{-1-1}$$

$$= -y^{-2}$$

$$\frac{dx}{dy} = \frac{-1}{y^2}$$

7. Find $\frac{dx}{dy}$ if $y = x^{\frac{2}{3}}$

Solution

$$y = x^{\frac{2}{3}}$$

Raise both sides to the exponent $\frac{3}{2}$ i.e. the inverse of $\frac{2}{3}$. This gives:

$$y^{\frac{3}{2}} = (x^{\frac{2}{3}})^{\frac{3}{2}}$$

$$y^{\frac{3}{2}} = x \quad \text{(Note that } \frac{2}{3} \times \frac{3}{2} = 1, \text{ and } x^1 = x)$$

$$x = y^{\frac{3}{2}}$$

$$\frac{dx}{dy} = \frac{3}{2} \times y^{\frac{3}{2}-1}$$

$$= \frac{3}{2} y^{\frac{1}{2}}$$

$$\frac{dx}{dy} = \frac{3\sqrt{y}}{2}$$

8. If $y = (5x + 7)^{\frac{3}{10}}$ find $\frac{dx}{dy}$.

Solution

$$y = (5x + 7)^{\frac{3}{10}}$$

214

Raise both sides to the exponent $\frac{10}{3}$ i.e. the inverse of $\frac{3}{10}$. This gives:

$$y^{\frac{10}{3}} = [(5x + 7)^{\frac{3}{10}}]^{\frac{10}{3}}$$

$$y^{\frac{10}{3}} = 5x + 7 \quad \text{(Note that } \frac{3}{10} \times \frac{10}{3} = 1\text{)}$$

$$y^{\frac{10}{3}} - 7 = 5x$$

$$x = \frac{y^{\frac{10}{3}} - 7}{5}$$

$$= \frac{y^{\frac{10}{3}}}{5} - \frac{7}{5}$$

$$x = \frac{1}{5}y^{\frac{10}{3}} - \frac{7}{5}$$

$$\frac{dx}{dy} = \frac{10}{3} \times \frac{1}{5}y^{\frac{10}{3} - 1}$$

$$= \frac{2}{3}y^{\frac{7}{3}}$$

$$\frac{dx}{dy} = \frac{2\sqrt[3]{y^7}}{3}$$

Exercise 16

1. Find $\frac{dx}{dy}$ if $y = 7x^3$

2. If $y = \sqrt[3]{1 + x^2}$ find $\frac{dx}{dy}$

3. If $y = 2x^5 - 3$. Find $\frac{dx}{dy}$

4. If $y = \frac{2}{3}x^3 - 9$, find $\frac{dx}{dy}$

5. If $y = \frac{2x + 1}{x}$, find $\frac{dx}{dy}$.

6. Find $\frac{dx}{dy}$ if $y = \frac{3}{x^2 - 5}$

7. Find $\frac{dx}{dy}$ if $y = 2x^{\frac{1}{4}}$

8. If $y = (x - 3)^{\frac{1}{5}}$ find $\frac{dx}{dy}$.

9. If $y = \dfrac{4x + 5}{x}$, find $\dfrac{dx}{dy}$.

10. Find $\dfrac{dx}{dy}$ if $y = \dfrac{1}{x^3 + 8}$

CHAPTER 17
DERIVATIVES OF INVERSE TRIGONOMETRIC FUNCTIONS

Recall that if $\sin x = y$, then $x = \sin^{-1} y$. This is referred to as inverse trigonometric function. The derivatives of inverse trigonometric functions are given below.

If $y = \sin^{-1} x$, then $\dfrac{dy}{dx} = \dfrac{1}{\sqrt{1-x^2}}$

If $y = \cos^{-1} x$, then $\dfrac{dy}{dx} = \dfrac{-1}{\sqrt{1-x^2}}$

If $y = \tan^{-1} x$, then $\dfrac{dy}{dx} = \dfrac{1}{1+x^2}$

If $y = \cot^{-1} x$, then $\dfrac{dy}{dx} = \dfrac{-1}{1+x^2}$

If $y = \sec^{-1} x$, then $\dfrac{dy}{dx} = \dfrac{1}{x\sqrt{x^2-1}}$

If $y = \csc^{-1} x$, then $\dfrac{dy}{dx} = \dfrac{-1}{x\sqrt{x^2-1}}$

Note that $\sin^{-1} x$ can also be written as $\arcsin x$. Other inverse function can be written in a similar way.

Examples

1. If $y = \cos^{-1} x$ find the $\dfrac{dy}{dx}$.

Solution

$$y = \cos^{-1} x$$

$$\frac{dy}{dx} = \frac{-1}{\sqrt{1-x^2}}$$

2. Find the derivative of $\sin^{-1} 3x$

Solution

$$y = \sin^{-1} 3x$$

Let us use the chain rule to solve this problem

Let $u = 3x$

Hence, $y = \sin^{-1} u$

$$\frac{du}{dx} = 3$$

$$\frac{dy}{du} = \frac{1}{\sqrt{1-u^2}}$$

Therefore, $\dfrac{dy}{dx} = \dfrac{dy}{du} \times \dfrac{du}{dx}$ \qquad (Chain rule)

$$= \frac{1}{\sqrt{1-u^2}} \times 3$$

$$= \frac{3}{\sqrt{1-u^2}}$$

$$= \frac{3}{\sqrt{1-(3x)^2}} \qquad \text{(Since } u = 3x\text{)}$$

$$\frac{dy}{dx} = \frac{3}{\sqrt{1-9x^2}}$$

3. Find the derivative of $\cot^{-1}x^2$

Solution

$$y = \cot^{-1}x^2$$

Let $u = x^2$

Hence, $y = \cot^{-1}u$

$$\frac{du}{dx} = 2x$$

$$\frac{dy}{du} = \frac{-1}{1+u^2}$$

Therefore, $\dfrac{dy}{dx} = \dfrac{dy}{du} \times \dfrac{du}{dx}$

$$= \frac{-1}{1+u^2} \times 2x$$

$$= \frac{-2x}{1+u^2}$$

$$= \frac{-2x}{1+(x^2)^2} \qquad \text{(since } u = x^2\text{)}$$

$$\frac{dy}{dx} = \frac{-2x}{1+x^4}$$

4. If $y = \sec^{-1}2x^3$, find $y = \dfrac{dy}{dx}$

Solution

$$y = \sec^{-1}2x^3$$

Let $u = 2x^3$

Hence, $y = \sec^{-1}u$

$$\frac{du}{dx} = 6x^2$$

$$\frac{dy}{du} = \frac{1}{u\sqrt{u^2-1}}$$

Therefore, $\dfrac{dy}{dx} = \dfrac{dy}{du} \times \dfrac{du}{dx}$

$$= \frac{1}{u\sqrt{u^2-1}} \times 6x^2$$

$$= \frac{6x^2}{u\sqrt{u^2-1}}$$

$$= \frac{6x^2}{2x^3\sqrt{(2x^3)^2-1}} \qquad \text{(since } u = 2x^3\text{)}$$

$$\frac{dy}{dx} = \frac{3}{x\sqrt{4x^6-1}} \qquad \text{(Note that } \frac{6x^2}{2x^3} = \frac{3}{x}\text{)}$$

5. Find the derivative $x^2\tan^{-1}x$

Solution

$$y = x^2\tan^{-1}x$$

we apply the product rule as follows:

$$u = x^2$$

$$v = \tan^{-1}x$$

$$\frac{du}{dx} = 2x$$

$$\frac{dv}{dx} = \frac{1}{1+x^2}$$

$$\frac{dy}{dx} = u\frac{dv}{dx} + v\frac{du}{dx} \qquad \text{(product rule)}$$

$$= x^2\left(\frac{1}{1+x^2}\right) + \tan^{-1}x(2x)$$

$$\frac{dy}{dx} = \frac{x^2}{1+x^2} + 2x\tan^{-1}x$$

6. If $y = 3x - 1 \csc^{-1}x^3$, find $\frac{dy}{dx}$.

Solution

$$y = 3x - 1 \csc^{-1}x^3$$

$$u = 3x - 1$$

$$\frac{du}{dx} = 3$$

$$v = \csc^{-1}x^3$$

$$\frac{dv}{dx} = \frac{-1}{x^3\sqrt{(x^3)^2-1}} \times 3x^2 \qquad \text{(From use of chain rule. Note that } 3x^2 \text{ is from the derivative of } x^3\text{)}$$

$$= \frac{-3x^2}{x^3\sqrt{x^6-1}}$$

$$\frac{dv}{dx} = \frac{-3}{x\sqrt{x^6-1}}$$

$$\frac{dy}{dx} = u\frac{dv}{dx} + v\frac{du}{dx} \qquad \text{(product rule)}$$

$$= 3x - 1 \left(\frac{-3}{x\sqrt{x^6 - 1}} \right) + \text{cosec}^{-1}x^3(3)$$

$$\frac{dy}{dx} = \frac{-3(3x - 1)}{x\sqrt{x^6 - 1}} + 3\text{cosec}^{-1}x^3$$

7. Find the derivative of $\cos^{-1}(5x - 3)$

<u>Solution</u>

$$y = \cos^{-1}(5x - 3)$$

Let $u = 5x - 3$

Hence, $y = \cos^{-1}u$

$$\frac{du}{dx} = 5$$

$$\frac{dy}{du} = \frac{-1}{\sqrt{1 - u^2}}$$

Therefore, $\dfrac{dy}{dx} = \dfrac{dy}{du} \times \dfrac{du}{dx}$

$$= \frac{-1}{\sqrt{1 - u^2}} \times 5$$

$$= \frac{-5}{\sqrt{1 - u^2}}$$

$$= \frac{-5}{\sqrt{1 - (5x - 3)^2}} \qquad \text{(Since } u = 5x - 3)$$

Expanding the bracket gives:

$$\frac{dy}{dx} = \frac{-5}{\sqrt{1 - (25x^2 - 15x - 15x + 9)}}$$

$$= \frac{-5}{\sqrt{1 - 25x^2 + 15x + 15x - 9}}$$

$$\frac{dy}{dx} = \frac{-5}{\sqrt{-25x^2 + 30x - 8}}$$

8. If $y = \tan^{-1}\left(\dfrac{1}{m}\right)$, find $\dfrac{dy}{dm}$

<u>Solution</u>

$$y = \tan^{-1}\left(\frac{1}{m}\right)$$

Let $u = \dfrac{1}{m} = m^{-1}$

Hence, $y = \tan^{-1}u$

$$\frac{du}{dm} = -1m^{-2}$$

$$\frac{du}{dm} = \frac{-1}{m^2}$$

$$\frac{dy}{du} = \frac{1}{1 + u^2}$$

Therefore, $\dfrac{dy}{dm} = \dfrac{dy}{du} \times \dfrac{du}{dm}$

$= \dfrac{1}{1+u^2} \times \dfrac{-1}{m^2}$

$= \dfrac{1}{1+\left(\frac{1}{m}\right)^2} \times \dfrac{-1}{m^2}$

$= \dfrac{-1}{m^2\left[1+\left(\frac{1}{m}\right)^2\right]}$

$= \dfrac{-1}{m^2\left(1+\frac{1}{m^2}\right)}$

$\dfrac{dy}{dm} = \dfrac{-1}{m^2+1}$

9. If y = sin3x, find $\dfrac{dx}{dy}$.

Solution

$\quad y = \sin 3x$

$\quad \sin^{-1}y = 3x \qquad$ (Recall that if sina = b, then a = sin⁻¹b)

$\quad x = \dfrac{\sin^{-1}y}{3}$

$\quad x = \dfrac{1}{3}\sin^{-1}y$

Hence, $\dfrac{dx}{dy} = \dfrac{1}{3}\dfrac{1}{\sqrt{1-y^2}}$

$\dfrac{dx}{dy} = \dfrac{1}{3\sqrt{1-y^2}}$

Note that this example asked us to find $\dfrac{dx}{dy}$ and not $\dfrac{dy}{dx}$

10. If y = (cos5x)² find $\dfrac{dx}{dy}$.

Solution

$\quad y = (\cos 5x)^2$

This can also be written as cos² 5x. Hence:

$\quad y = \cos^2 5x.$

Or, $\cos^2 5x = y$

taking the square root of both sides gives:

$\quad \cos 5x = \sqrt{y}$

$\quad 5x = \cos^{-1}\sqrt{y}$

221

$$x = \frac{\cos^{-1}\sqrt{y}}{5}$$

$$= \frac{1}{5}\cos^{-1}\sqrt{y}$$

$$x = \frac{1}{5}\cos^{-1}y^{\frac{1}{2}}$$

Hence, $\dfrac{dx}{dy} = \dfrac{1}{5} \cdot \dfrac{-1}{\sqrt{1 - (y^{\frac{1}{2}})^2}} \times \dfrac{1}{2}y^{-\frac{1}{2}}$ (Note that $\dfrac{1}{2}y^{-\frac{1}{2}}$ is from the derivative of $y^{\frac{1}{2}}$)

$$= \frac{-1}{5\sqrt{1 - y}} \times \frac{1}{2y^{\frac{1}{2}}}$$

$$= \frac{-1}{10y^{\frac{1}{2}}\sqrt{1 - y}}$$

$$= \frac{-1}{10\sqrt{y}\sqrt{1 - y}}$$

$$\frac{dx}{dy} = \frac{-1}{10\sqrt{y(1 - y)}}$$

Exercise 17

1. If $y = \sin^{-1}2x$ find the $\dfrac{dy}{dx}$.

2. Find the derivative of $\cos^{-1}x$

3. Find the derivative of $\cot^{-1}3x^2$

4. If $y = \cot^{-1}x^4$, find $y = \dfrac{dy}{dx}$

5. Find the derivative $3x\sec^{-1}x$

6. If $y = x^2 + 3\tan^{-1}x$, find $\dfrac{dy}{dx}$.

7. Find the derivative of $\tan^{-1}(x + 5)$

8. If $y = \cos^{-1}\left(\dfrac{1}{x}\right)$, find $\dfrac{dy}{dx}$

9. If $y = \cos5x$, find $\dfrac{dx}{dy}$.

10. If $y = (\sec x)^2$ find $\dfrac{dx}{dy}$.

11. Find the derivative $5x\tan^{-1}3x$

12. If $y = x^2 + 4\sin^{-1}x^5$, find $\dfrac{dy}{dx}$.

13. Find the derivative of $\csc^{-1}x^3$

14. If $y = \tan 2x$, find $\dfrac{dx}{dy}$.

15. If $y = \sin^3 5x$, find $\dfrac{dx}{dy}$.

CHAPTER 18
DERIVATIVES OF HYPERBOLIC FUNCTIONS

Hyperbolic functions are functions in calculus which are expressed by the combination of the exponential functions e^x and e^{-x}. The derivatives of the six main hyperbolic functions are as given below.

1. If $y = \sinh x$, then $\dfrac{dy}{dx} = \cosh x$

2. If $y = \cosh x$, then $\dfrac{dy}{dx} = \sinh x$

3. If $y = \tanh x$, then $\dfrac{dy}{dx} = \text{sech}^2 x$

4. If $y = \coth x$, then $\dfrac{dy}{dx} = -\text{cosech}^2 x$

5. If $y = \text{sech} x$, then $\dfrac{dy}{dx} = -\text{sech} x \tanh x$

6. If $y = \text{cosech} x$, then $\dfrac{dy}{dx} = -\text{cosech} x \coth x$

Note that $\sinh x = \dfrac{e^x - e^{-x}}{2}$ and $\cosh x = \dfrac{e^x + e^{-x}}{2}$

Examples

1. If $y = \cosh x - 5\sinh x$, find $\dfrac{dy}{dx}$.

Solution

$$y = \cosh x - 5\sinh x$$

$$\frac{dy}{dx} = \sinh x - 5\cosh x$$

2. Find the derivative of $2x^3 \coth x$

Solution

$$y = 2x^3 \coth x$$

Using product rule gives $\dfrac{dy}{dx}$ as follows:

$$\frac{dy}{dx} = 2x^3\left[\frac{d(\coth x)}{dx}\right] + \coth x\left[\frac{d(2x^3)}{dx}\right]$$

$$= 2x^3(-\text{cosech}^2 x) + \coth x(6x^2)$$

$$= -2x^3\text{cosech}^2 x + 6x^2\coth x$$

$$\frac{dy}{dx} = -2x^2(x\text{cosech}^2 x - 3\coth x)$$

Take note of the change in sign of the term in the bracket. This is due to the negative sign outside the bracket. Expanding the bracket gives the original expression that was factorized.

3. If $y = \dfrac{\cosh x}{x^2 + 1}$ find $\dfrac{dy}{dx}$.

Solution

Let us use product rule to obtained $\dfrac{dy}{dx}$ as follows:

$$\frac{dy}{dx} = \frac{(x^2+1)\left[\frac{d(\cosh x)}{dx}\right] - \cosh x\left[\frac{d(x^2+1)}{dx}\right]}{(x^2+1)^2}$$

$$= \frac{(x^2+1)(\sinh x) - \cosh x(2x)}{(x^2+1)^2}$$

$$\frac{dy}{dx} = \frac{\sinh x(x^2+1) - 2x\cosh x}{(x^2+1)^2}$$

4. Find the derivative of $(\sinh 3x)^2$

Solution

$y = (\sinh 3x)^2$ (This can also be written as $\sinh^2 3x$)

Let $u = \sinh 3x$

Hence, $y = u^2$

$\dfrac{du}{dx} = 3\cosh 3x$ (Note that 3 is from the derivative of $3x$)

$\dfrac{dy}{du} = 2u$

$\dfrac{dy}{dx} = \dfrac{dy}{du} \times \dfrac{du}{dx}$

$\quad = 2u \times 3\cosh 3x$

$\quad = 6u\cosh 3x$

$\quad = 6\sinh 3x \cosh 3x$ (u has been replaced with $\sinh 3x$)

5. Find the derivative of $\sinh^4 2x^3$

Solution

$y = \sinh^4 2x^3$

Let $u = \sinh 2x^3$

Hence, $y = u^4$

$\dfrac{du}{dx} = 6x^2\cosh 2x^3$ (Note that $6x^2$ is from the derivative of $2x^3$)

$\dfrac{dy}{du} = 4u^3$

$\dfrac{dy}{dx} = \dfrac{dy}{du} \times \dfrac{du}{dx}$

$\quad = 4u^3 \times 6x^2\cosh 2x^3$

$\quad = 24x^2 u^3\cosh 2x^3$

$\dfrac{dy}{dx} = 24x^2\sinh^3 2x^3\cosh 2x^3$ (u has been replaced with $\sinh 2x^3$)

6. Find the derivative of $\text{cosech}4x^3$

Solution

$y = \text{cosech}4x^3$

$\dfrac{dy}{dx} = -\text{cosech}4x^3\text{coth}4x^3 \times \dfrac{d(4x^3)}{dx}$　　　(Note that the derivative of cosech is $-$cosechcoth)

$\quad\quad = -\text{cosech}4x^3\text{coth}4x^3 \times 12x^2$

$\quad\quad = -12x^2\text{cosech}4x^3\text{coth}4x^3$

7. Find the derivative of $\text{coth}^2 2x^4$

Solution

$y = \text{coth}^2 2x^4$

Let $u = \text{coth}2x^4$

Hence, $y = u^2$

$\dfrac{du}{dx} = -\text{cosech}^2 2x^4 \times \dfrac{d(2x^4)}{dx}$

$\quad\quad = -\text{cosech}^2 2x^4 \times 8x^3$

$\dfrac{du}{dx} = -8x^3\text{cosech}^2 2x^4$

$\dfrac{dy}{du} = 2u$

Therefore, $\dfrac{dy}{dx} = \dfrac{dy}{du} \times \dfrac{du}{dx}$

$\quad\quad = 2u \times (-8x^3\text{cosech}^2 2x^4)$

$\quad\quad = -16x^3 u\,\text{cosech}^2 2x^4$

$\dfrac{dy}{dx} = -16x^3\text{coth}2x^4\text{cosech}^2 2x^4$　　　(Since $u = \text{coth}2x^4$)

8. Find the derivative of $5x\text{sinh}2x$

Solution

$y = 5x\text{sinh}2x$

We apply the product rule of differentiation as follows:

Let $u = 5x$

and $v = \text{sinh}2x$

$\dfrac{du}{dx} = 5$

$\dfrac{dv}{dx} = 2\text{cosh}2x$　　　(Note that the differentiation of $2x$ gives 2)

Hence, $\dfrac{dy}{dx} = u\dfrac{dv}{dx} + v\dfrac{du}{dx}$

$\quad\quad = (5x \times 2\text{cosh}2x) + (\text{sinh}2x \times 5)$

$\quad\quad = 10x\text{cosh}2x + 5\text{sinh}2x$

$$\frac{dy}{dx} = 5(2x\cosh 2x + \sinh 2x)$$

9. If $y = \dfrac{1 + \cosh 2x}{\sinh 2x}$ find $\dfrac{dy}{dx}$.

Solution

$$y = \frac{1 + \cosh 2x}{\sinh 2x}$$

We apply quotient rule as follows:

$u = 1 + \cosh 2x$

$v = \sinh 2x$

$\dfrac{du}{dx} = 2\sinh 2x$

$\dfrac{dv}{dx} = 2\cosh 2x$

Hence, $\dfrac{dy}{dx} = \dfrac{v\frac{du}{dx} - u\frac{dv}{dx}}{v^2}$ (Quotient rule)

$$= \frac{\sinh 2x(2\sinh 2x) - (1 + \cosh 2x)(2\cosh 2x)}{(\sinh 2x)^2}$$

$$= \frac{2\sin^2 2x - (2\cosh 2x + 2\cosh^2 2x)}{\sinh^2 2x} \quad \text{(Note that } (\sinh 2x)^2 = \sinh^2 2x)$$

$$= \frac{2\sinh^2 2x - 2\cosh 2x - 2\cosh^2 2x}{\sinh^2 2x}$$

$$= \frac{2\sinh^2 2x - 2\cosh^2 2x - 2\cosh 2x}{\sinh^2 2x}$$

$$= \frac{-2(\cosh^2 2x - \sinh^2 2x) - 2\cosh 2x}{\sinh^2 2x}$$

$$= \frac{-2(1) - 2\cosh 2x}{\sinh^2 2x} \quad \text{(Note that } \cosh^2 2x - \sinh^2 2x = 1)$$

$$\frac{dy}{dx} = \frac{-2(1 + \cosh 2x)}{\sinh^2 2x}$$

Or, $\dfrac{dy}{dx} = \dfrac{-2(1 + \cosh 2x)}{\cosh^2 2x - 1}$ (Note that since $\cosh^2 2x - \sinh^2 2x = 1$, then $\cosh^2 2x - 1 = \sinh^2 2x$)

$$= \frac{-2(1 + \cosh 2x)}{(\cosh 2x + 1)(\cosh 2x - 1)}$$

Note that from difference of two squares we have: $a^2 - b^2 = (a + b)(a - b)$.

Hence, $\cosh^2 2x - 1$ is also a difference of two squares since 1 is also 1^2. Therefore, $\cosh^2 2x - 1 = (\cosh 2x + 1)(\cosh 2x - 1)$ as represented above. Hence the expression above becomes:

$$\frac{dy}{dx} = \frac{-2}{(\cosh 2x - 1)}$$

10. Differentiate $y = \dfrac{\sinh^2 x}{x}$

Solution

$$y = \frac{\sinh^2 x}{x}$$

We use quotient rule as follows:

Therefore, $u = \sinh^2 x$

$$v = x$$

Hence, $\dfrac{du}{dx} = 2\sinh x \cosh x$

Since, $v = x$

Then, $\dfrac{dv}{dx} = 1$

Hence, $\dfrac{dy}{dx} = \dfrac{v\frac{du}{dx} - u\frac{dv}{dx}}{v^2}$

$$= \frac{x(2\sinh x\cosh x) - \sinh^2 x(1)}{x^2}$$

$$= \frac{2x\sinh x\cosh x - \sin^2 x}{x^2}$$

$$= \frac{\sinh x(2x\cosh x - \sinh x)}{x^2}$$

Exercise 18

1. If $y = \sinh 3x - 2\cosh x$, find $\dfrac{dy}{dx}$.

2. Find the derivative of $5x^2 \operatorname{sech} x$

3. If $y = \dfrac{\sinh 2x}{3x^2}$ find $\dfrac{dy}{dx}$.

4. Find the derivative of $\cosh^2 3x$

5. Find the derivative of $\cosh^3 4x^5$

6. Find the derivative of $\coth 2x^5$

7. Find the derivative of $(\operatorname{sech} 2x^2)^3$

8. Find the derivative of $x^2 \cosh 5x$

9. If $y = \dfrac{\tanh 5x}{\coth 3x}$ find $\dfrac{dy}{dx}$.

10. Differentiate $y = \dfrac{2\cosh^3 x}{3x}$

CHAPTER 19
DERIVATIVE OF LOGARITHMIC FUNCTIONS

If $y = \log_a x$, then $\dfrac{dy}{dx} = \dfrac{1}{x}\log_a e$, where a is any base.

If $y = \log_e x$, then $\dfrac{dy}{dx} = \dfrac{1}{x}$

Note that $\log_e x$ can also be represented as $\ln x$ and the value of e is 2.718 (to 3 decimal places)

Examples

1. Find the derivative of $\log_a 2x$.

Solution

$\quad y = \log_a 2x$

We will use chain rule since we have a function that is not just x but $2x$.

Let $u = 2x$

Therefore, $y = \log_a u$

$\dfrac{du}{dx} = 2$

$\dfrac{dy}{du} = \dfrac{1}{u}\log_a e$

$\dfrac{dy}{dx} = \dfrac{dy}{du} \times \dfrac{du}{dx}$

$\quad\quad = \dfrac{1}{u}\log_a e \times 2$

$\quad\quad = \dfrac{2}{u}\log_a e$

$\quad\quad = \dfrac{2}{2x}\log_a e \quad\quad$ (since u = 2x)

$\dfrac{dy}{dx} = \dfrac{1}{x}\log_a e$

2. Find the derivative of $\log_a(5x - 1)$

Solution

$\quad y = \log_a(5x - 1)$

Let $u = 5x - 1$

Therefore, $y = \log_a u$

$\dfrac{du}{dx} = 5$

$\dfrac{dy}{du} = \dfrac{1}{u}\log_a e$

$\dfrac{dy}{dx} = \dfrac{dy}{du} \times \dfrac{du}{dx}$

$$= \frac{1}{u}\log_a e \times 5$$

$$= \frac{5}{u}\log_a e$$

$$\frac{dy}{dx} = \frac{5}{5x-1}\log_a e \qquad (\text{since } u = 5x - 1)$$

3. Differentiate $\log_a(4x - 3)^2$ with respect to x

Solution

$$y = \log_a(4x - 3)^2$$

Let $u = (4x - 3)^2$

Therefore, $y = \log_a u$

$$\frac{du}{dx} = 2(4x - 3)^{2-1} \times 4 \qquad (\text{Note that 4 is from the derivative of } 4x - 3)$$

$$\frac{du}{dx} = 8(4x - 3)$$

$$\frac{dy}{du} = \frac{1}{u}\log_a e$$

$$\frac{dy}{dx} = \frac{dy}{du} \times \frac{du}{dx}$$

$$= \frac{1}{u}\log_a e \times 8(4x - 3)$$

$$= \frac{8(4x-3)}{u}\log_a e$$

$$= \frac{8(4x-3)}{(4x-3)^2}\log_a e \qquad [\text{since } u = (4x - 3)^2]$$

$$\frac{dy}{dx} = \frac{8}{4x-3}\log_a e \qquad (\text{One } 4x - 3 \text{ cancels out})$$

4. If $y = \log_a\sqrt{1 + 2x}$, find $\frac{dy}{dx}$

Solution

$$y = \log_a\sqrt{1 + 2x}$$

Let $u = \sqrt{1 + 2x} = (1 + 2x)^{\frac{1}{2}}$

Therefore, $y = \log_a u$

$$\frac{du}{dx} = \frac{1}{2}(1 + 2x)^{\frac{1}{2}-1} \times 2$$

$$= (1 + 2x)^{-\frac{1}{2}} \qquad (\text{Note that } \frac{1}{2} \times 2 = 1)$$

$$\frac{du}{dx} = \frac{1}{(1+2x)^{\frac{1}{2}}}$$

$$\frac{dy}{du} = \frac{1}{u}\log_a e$$

$$\frac{dy}{dx} = \frac{dy}{du} \times \frac{du}{dx}$$

$$= \frac{1}{u} \log_a e \times \frac{1}{(1+2x)^{\frac{1}{2}}}$$

$$= \frac{1}{(1+2x)^{\frac{1}{2}}} \log_a e \times \frac{1}{(1+2x)^{\frac{1}{2}}} \qquad \text{[Note that u has been replaced with } (1+2x)^{\frac{1}{2}}]$$

$$\frac{dy}{dx} = \frac{1}{1+2x} \log_a e \qquad \text{[Note that } (1+2x)^{\frac{1}{2}} \times (1+2x)^{\frac{1}{2}} = (1+2x)^1 \text{ by adding exponents]}$$

5. Find $\dfrac{dy}{dx}$ given that y = $\log_a \dfrac{1-3x}{1+3x}$

Solution

$$y = \log_a \frac{1-3x}{1+3x}$$

Or, y = $\log_a(1 - 3x) - \log_a(1 + 3x)$ $\qquad$ (Recall that $\log_x\left(\dfrac{a}{b}\right) = \log_x a - \log_x b$)

$$\frac{dy}{dx} = \frac{-3}{1-3x} \log_a e - \frac{3}{1+3x} \log_a e$$

$$= -\log_a e \left(\frac{3}{1-3x} + \frac{3}{1+3x}\right)$$

$$= -\log_a e \left[\frac{3(1+3x) + 3(1-3x)}{(1-3x)(1+3x)}\right]$$

$$= -\log_a e \left[\frac{3 + 9x + 3 - 9x}{1 - 9x^2}\right] \qquad \text{[Note that } (1-3x)(1+3x) = 1 - 9x^2]$$

$$= -\log_a e \left(\frac{6}{1 - 9x^2}\right)$$

$$\frac{dy}{dx} = \frac{-6}{1 - 9x^2} \log_a e$$

6. If y = $\log_{10}(x^2 - 2)$, find $\dfrac{dy}{dx}$.

Solution

$$y = \log_{10}(x^2 - 2)$$

The value 10 represents a in other examples. So we are going to differentiate y like the examples above except that we will write 10 wherever 'a' should be.

$$y = \log_{10}(x^2 - 2)$$

$$u = x^2 - 2$$

Hence, y = $\log_{10} u$

$$\frac{du}{dx} = 2x$$

$$\frac{dy}{du} = \frac{1}{u} \log_{10} e \qquad \text{(Just like differentiating } \log_a u)$$

$$\frac{dy}{dx} = \frac{dy}{du} \times \frac{du}{dx}$$

$$= \frac{1}{u} \log_{10}e \times 2x$$

$$= \frac{2x}{u} \log_{10}e$$

$$\frac{dy}{dx} = \frac{2x}{x^2 - 2} \log_{10}e$$

7. Find $\frac{dy}{dx}$ if $y = \log_{10}\frac{1}{x}$

<u>Solution</u>

$$y = \log_{10}\frac{1}{x}$$

Let us solve this question directly without using $u = \frac{1}{x}$ as follows:

$$\frac{dy}{dx} = \frac{\frac{d\left(\frac{1}{x}\right)}{dx}}{\frac{1}{x}} \times \log_{10}e$$

$$= \frac{\frac{d\left(x^{-1}\right)}{dx}}{x^{-1}} \times \log_{10}e$$

$$= \frac{-x^{-2}}{x^{-1}} \times \log_{10}e \qquad (\text{Note that the derivative of } x^{-1} \text{ is } -x^{-2})$$

$$= -x^{-2} \times x^1 \times \log_{10}e \qquad (\text{Since } \frac{1}{x^{-1}} = x^1)$$

$$= -x^{-1} \times \log_{10}e \qquad (\text{After adding the exponents of } x)$$

$$\frac{dy}{dx} = -\frac{1}{x} \log_{10}e$$

8. Find the derivative of $\log_e(2 - x^3)$

<u>Solution</u>

$$y = \log_e(2 - x^3)$$

Note that the base here is 'e' and not 'a'.

Let $u = 2 - x^3$

Hence, $y = \log_e u$

$$\frac{du}{dx} = -3x^2$$

$$\frac{dy}{du} = \frac{1}{u} \qquad (\text{Recall that the derivative of } \log_e x = \frac{1}{x})$$

$$\frac{dy}{dx} = \frac{dy}{du} \times \frac{du}{dx}$$

$$= \frac{1}{u} \times -3x^2$$

$$= \frac{-3x^2}{u}$$

$$\frac{dy}{dx} = \frac{-3x^2}{2-x^3}$$

9. Find the derivative of $(\log_e 5x)^2$

Solution

$$y = (\log_e 5x)^2$$

Let $u = \log_e 5x$

Hence, $y = u^2$

$$\frac{du}{dx} = \frac{\frac{d(5x)}{dx}}{5x}$$

$$= \frac{5}{5x}$$

$$\frac{du}{dx} = \frac{1}{x}$$

$$\frac{dy}{du} = 2u$$

$$\frac{dy}{dx} = \frac{dy}{du} \times \frac{du}{dx}$$

$$= 2u \times \frac{1}{x}$$

$$= \frac{2u}{x}$$

$$\frac{dy}{dx} = \frac{2}{x}\log_e 5x \quad \text{(Since } u = \log_e 5x\text{)}$$

10. If $y = \ln\sqrt{3x^2 - 4}$, find $\frac{dy}{dx}$.

Solution

$$y = \ln\sqrt{3x^2 - 4}$$

Note that $\ln\sqrt{3x^2 - 4}$ is also the same as $\log_e\sqrt{3x^2 - 4}$

Hence, $y = \log_e\sqrt{3x^2 - 4}$

Or, $y = \log_e(3x^2 - 4)^{\frac{1}{2}}$

Let $u = (3x^2 - 4)^{\frac{1}{2}}$

Hence, $y = \log_e u$

$$\frac{du}{dx} = \frac{1}{2}(3x^2 - 4)^{\frac{1}{2}-1} \times 6x \quad \text{(Note that } 6x \text{ is from the derivative of } 3x^2 - 4\text{)}$$

$$= \frac{6x}{2}(3x^2 - 4)^{-\frac{1}{2}}$$

$$\frac{du}{dx} = \frac{3x}{(3x^2-4)^{\frac{1}{2}}}$$

$$\frac{dy}{du} = \frac{1}{u}$$

$$\frac{dy}{dx} = \frac{dy}{du} \times \frac{du}{dx}$$

$$= \frac{1}{u} \times \frac{3x}{(3x^2 - 4)^{\frac{1}{2}}}$$

$$= \frac{1}{(3x^2 - 4)^{\frac{1}{2}}} \times \frac{3x}{(3x^2 - 4)^{\frac{1}{2}}}$$

$$\frac{dy}{dx} = \frac{3x}{3x^2 - 4} \quad \text{(Note that } (3x^2 - 4)^{\frac{1}{2}} \times (3x^2 - 4)^{\frac{1}{2}} = 3x^2 - 4, \text{ after adding their exponents)}$$

11. Find $\frac{dy}{dx}$ if $y = \sqrt[3]{x}\, \ln 2x$.

Solution

$y = \sqrt[3]{x}\, \ln 2x$

We are going to apply product rule here.

$u = \sqrt[3]{x} = x^{\frac{1}{3}}$

$v = \ln 2x$ (Note that $\ln 2x$ is the same as $\log_e 2x$)

$$\frac{du}{dx} = \frac{1}{3}x^{\frac{1}{3} - 1}$$

$$= \frac{1}{3}x^{-\frac{2}{3}}$$

$$\frac{dv}{dx} = \frac{2}{2x}$$

$$= \frac{1}{x}$$

Hence, $\frac{dy}{dx} = u\frac{dv}{dx} + v\frac{du}{dx}$ (Product rule)

$$= x^{\frac{1}{3}}\left(\frac{1}{x}\right) + \ln 2x \left(\frac{1}{3}x^{-\frac{2}{3}}\right)$$

$$= x^{\frac{1}{3}}\left(x^{-1}\right) + \frac{1}{3}x^{-\frac{2}{3}} \ln 2x$$

$$= x^{-\frac{2}{3}}\left(1 + \frac{\ln 2x}{3}\right)$$

$$= x^{-\frac{2}{3}}\left(\frac{3 + \ln 2x}{3}\right)$$

$$\frac{dy}{dx} = \frac{1}{x^{\frac{2}{3}}}\left(\frac{3 + \ln 2x}{3}\right)$$

Or, $\frac{dy}{dx} = \frac{1}{\sqrt[3]{x^2}}\left(\frac{3 + \ln 2x}{3}\right)$

12. Find $\frac{dy}{dx}$ given that $y = \ln(1 + 2x)^2$

Solution

$y = \ln(1 + 2x)^2$ [Also $y = \log_e(1 + 2x)^2$]

Let $u = (1 + 2x)^2$

Hence, $y = \ln u$

$$\frac{du}{dx} = 2(1 + 2x) \times \frac{d(2x)}{dx}$$

$$= 2(1 + 2x) \times 2x$$

$$\frac{du}{dx} = 4(1 + 2x)$$

$$\frac{dy}{du} = \frac{1}{u}$$

$$\frac{dy}{dx} = \frac{1}{u} \times 4(1 + 2x)$$

$$= \frac{4(1+2x)}{u}$$

$$= \frac{4(1+2x)}{(1+2x)^2}$$

$$= \frac{4}{1+2x}$$

13. Differentiate with respect to x: $5x\ln(3x^2 - 2)$

Solution

$y = 5x\ln(3x^2 - 2)$

By using product rule:

$u = 5x$

$v = \ln(3x^2 - 2)$

$$\frac{du}{dx} = 5$$

$$\frac{dv}{dx} = \frac{6x}{3x^2 - 2}$$

Hence, $\dfrac{dy}{dx} = u\dfrac{dv}{dx} + v\dfrac{du}{dx}$

$$= 5x\left(\frac{6x}{3x^2 - 2}\right) + \ln(3x^2 - 2)(5)$$

$$\frac{dy}{dx} = \frac{30x^2}{3x^2 - 2} + 5\ln(3x^2 - 2)$$

14. Find the derivative of $\dfrac{\ln x}{x}$

Solution

$y = \dfrac{\ln x}{x}$

We are going to use quotient rule as follows:

$u = \ln x$

$v = x$

$$\frac{du}{dx} = \frac{1}{x}$$

$$\frac{dv}{dx} = 1$$

Hence, $\dfrac{dy}{dx} = \dfrac{v\dfrac{du}{dx} - u\dfrac{dv}{dx}}{v^2}$

$$= \frac{x\left(\frac{1}{x}\right) - \ln x \,(1)}{x^2}$$

$$\frac{dy}{dx} = \frac{1 - \ln x}{x^2}$$

Exercice 19

1. Find the derivative of $\log_a 7x$.

2. Find the derivative of $\log_a(x^3 + 5)$

3. Differentiate $\log_a(2x^3 - 5)^6$ with respect to x

4. If $y = \log_a x^{\frac{2}{3}}$, find $\dfrac{dy}{dx}$

5. Find $\dfrac{dy}{dx}$ given that $y = \log_a \dfrac{1 + x^2}{1 - x^2}$

6. If $y = \log_5(5x^3 - 1)$, find $\dfrac{dy}{dx}$.

7. Find $\dfrac{dy}{dx}$ if $y = \log_2 \dfrac{1}{x^2}$

8. Find the derivative of $\log_e(x^2 + 3)$

9. Find the derivative of $(\log_e x)^3$

10. If $y = \ln\sqrt{1 - 2x^5}$, find $\dfrac{dy}{dx}$.

11. Find $\dfrac{dy}{dx}$ if $y = x^4 \ln x^2$

12. Find $\dfrac{dy}{dx}$ given that $y = \ln(3 - 7x)^4$

13. Differentiate with respect to x: $3x^2 \ln(4x^3 + 1)$

14. Find the derivative of $\dfrac{\ln x^2}{x^2}$

15. Find the derivative of $\log_e(1 - 5x^2)^3$

236

CHAPTER 20
DERIVATIVE OF EXPONENTIAL FUNCTIONS

If $y = a^x$, then $\dfrac{dy}{dx} = a^x \log_e a$, where 'a' is any number.

If $y = e^x$, then $\dfrac{dy}{dx} = e^x$

Examples

1. Differentiate with respect to x: a^{5x}

Solution

$y = a^{5x}$

Let $u = 5x$

$y = a^u$

$\dfrac{du}{dx} = 5$

$\dfrac{dy}{du} = a^u \log_e a$

Hence, $\dfrac{dy}{dx} = \dfrac{dy}{du} \times \dfrac{du}{dx}$

$= a^u \log_e a \times 5$

$= 5a^u \log_e a$

$\dfrac{dy}{dx} = 5a^{5x} \log_e a \qquad$ (since $u = 5x$)

2. Find the derivative of $a^{x^2 - 3x + 4}$

Solution

$y = a^{x^2 - 3x + 4}$

Let us solve this example directly without using $u = x^2 - 3x + 4$

$\dfrac{dy}{dx} = \dfrac{d(x^2 - 3x + 4)}{dx} \times a^{x^2 - 3x + 4} \times \log_e a$

$= (2x - 3)(a^{x^2 - 3x + 4} \times \log_e a)$

3. If $y = 3x^2 a^{5x}$, find $\dfrac{dy}{dx}$.

Solution

$y = 3x^2 a^{5x}$

We apply product rule as follows:

$u = 3x^2$

$v = a^{5x}$

$$\frac{du}{dx} = 6x$$

$$\frac{dv}{dx} = 5a^{5x}\log_e a$$

Hence, $\dfrac{dy}{dx} = u\dfrac{dv}{dx} + v\dfrac{du}{dx}$

$$= 3x^2(5a^{5x}\log_e a) + a^{5x}(6x)$$

$$= 15x^2 a^{5x}\log_e a + 6xa^{5x}$$

Factorizing the expression above gives:

$$\frac{dy}{dx} = 3xa^{5x}(5x\log_e a + 2)$$

4. Find the derivative of $\dfrac{e^x + e^{-x}}{e^x - e^{-x}}$

Solution

$$y = \frac{e^x + e^{-x}}{e^x - e^{-x}}$$

$$u = e^x + e^{-x}$$

$$v = e^x - e^{-x}$$

$$\frac{du}{dx} = e^x - e^{-x}$$ (The derivative of $e^{-x} = -1 \times e^{-x} = -e^{-x}$. The value -1 is from the derivative of $-x$)

$$\frac{dv}{dx} = e^x + e^{-x}$$

Hence, $\dfrac{dy}{dx} = \dfrac{v\dfrac{du}{dx} - u\dfrac{dv}{dx}}{v^2}$ (Quotient rule)

$$= \frac{(e^x - e^{-x})(e^x - e^{-x}) - (e^x + e^{-x})(e^x + e^{-x})}{(e^x - e^{-x})^2}$$

Expanding the numerator and denominator gives:

$$\frac{dy}{dx} = \frac{(e^x)(e^x) - (e^x)(e^{-x}) - (e^{-x})(e^x) + (e^{-x})(e^{-x}) - [(e^x)(e^x) + (e^x)(e^{-x}) + (e^{-x})(e^x) + (e^{-x})(e^{-x})]}{(e^x - e^{-x})^2}$$

$$= \frac{e^{2x} - 1 - 1 + e^{-2x} - (e^{2x} + 1 + 1 + e^{-2x})}{(e^x - e^{-x})^2}$$ (Note that exponents have been added, and $e^0 = 1$)

$$= \frac{e^{2x} - 2 + e^{-2x} - e^{2x} - 1 - 1 - e^{-2x})}{(e^x - e^{-x})^2}$$

$$\frac{dy}{dx} = \frac{-4}{(e^x - e^{-x})^2}$$ (e^{2x} and e^{-2x} cancel out)

5. If $y = e^{\sqrt{x}}$ find $\dfrac{dy}{dx}$.

Solution

$$y = e^{\sqrt{x}} = e^{x^{\frac{1}{2}}}$$

$$\frac{dy}{dx} = \frac{d(x)^{\frac{1}{2}}}{dx} \times e^{x^{\frac{1}{2}}}$$

$$= \frac{1}{2}x^{-\frac{1}{2}} \times e^{x^{\frac{1}{2}}}$$

$$= \frac{1}{2}\left(\frac{1}{x^{\frac{1}{2}}}\right) \times e^{x^{\frac{1}{2}}}$$

$$= \frac{e^{x^{\frac{1}{2}}}}{2x^{\frac{1}{2}}}$$

$$\frac{dy}{dx} = \frac{e^{\sqrt{x}}}{2\sqrt{x}}$$

6. Find $\frac{dy}{dx}$ if $y = x^{\frac{1}{2}} e^{x^{\frac{1}{2}}}$

Solution

$$y = x^{\frac{1}{2}} e^{x^{\frac{1}{2}}}$$

From product rule:

$$\frac{dy}{dx} = x^{\frac{1}{2}}\frac{d(e^{x^{\frac{1}{2}}})}{dx} + e^{x^{\frac{1}{2}}}\frac{d(x^{\frac{1}{2}})}{dx}$$

$$= x^{\frac{1}{2}}\left(\frac{e^{x^{\frac{1}{2}}}}{2x^{\frac{1}{2}}}\right) + e^{x^{\frac{1}{2}}}\left(\frac{1}{2}x^{-\frac{1}{2}}\right) \quad \text{(Note that the derivative of } e^{x^{\frac{1}{2}}} \text{ is } \frac{e^{x^{\frac{1}{2}}}}{2x^{\frac{1}{2}}} \text{ from example 5)}$$

$$= \frac{e^{x^{\frac{1}{2}}}}{2} + \frac{e^{x^{\frac{1}{2}}}}{2x^{\frac{1}{2}}}$$

$$= \frac{x^{\frac{1}{2}}e^{x^{\frac{1}{2}}} + e^{x^{\frac{1}{2}}}}{2x^{\frac{1}{2}}}$$

$$\frac{dy}{dx} = \frac{e^{x^{\frac{1}{2}}}(x^{\frac{1}{2}} + 1)}{2x^{\frac{1}{2}}}$$

7. Find the derivative of $a^{2x} - a^{-2x}$

Solution

$$y = a^{2x} - a^{-2x}$$

$$\frac{dy}{dx} = \frac{d(2x)}{dx}(a^{2x}\log_e a) - \left[\frac{d(-2x)}{dx}(a^{-2x}\log_e a)\right]$$

$$= 2a^{2x}\log_e a - (-2a^{-2x}\log_e a)$$

$$= 2a^{2x}\log_e a + 2a^{-2x}\log_e a$$

$$= 2\log_e a(a^{2x} + a^{-2x})$$

8. Find the derivative of $e^{2x}\log_e 3x$

Solution

$$y = e^{2x}\log_e 3x$$

By product rule:

$$u = e^{2x}$$

$$v = \log_e 3x \quad (\text{or } v = \ln 3x)$$

$$\frac{du}{dx} = 2e^{2x}$$

$$\frac{dv}{dx} = \frac{3}{3x}$$

$$= \frac{1}{x}$$

$$\frac{dy}{dx} = e^{2x}\left(\frac{1}{x}\right) + \log_e 3x(2e^{2x})$$

$$= \frac{e^{2x}}{x} + 2e^{2x}\log_e 3x$$

$$\frac{dy}{dx} = e^{2x}(\frac{1}{x} + 2\log_e 3x)$$

9. Differentiate with respect to x: $2\sqrt{xe^x}$

Solution

$$y = 2\sqrt{xe^x}$$

Or, $y = 2\sqrt{x}\,\sqrt{e^x}$

From product rule:

$$u = 2\sqrt{x} = 2x^{\frac{1}{2}}$$

$$v = \sqrt{e^x} = (e^x)^{\frac{1}{2}}$$

$$v = e^{\frac{1}{2}x} \quad \text{(When the exponents are multiplied)}$$

$$\frac{du}{dx} = \frac{1}{2} \times 2x^{-\frac{1}{2}}$$

$$= x^{-\frac{1}{2}}$$

$$\frac{dv}{dx} = \frac{1}{2}e^{\frac{1}{2}x} \quad \text{(Note that } \frac{1}{2} \text{ is from the derivative of } \frac{1}{2}x)$$

$$\frac{dy}{dx} = u\frac{dv}{dx} + v\frac{du}{dx}$$

$$= 2x^{\frac{1}{2}}(\frac{1}{2}e^{\frac{1}{2}x}) + e^{\frac{1}{2}x}(x^{-\frac{1}{2}})$$

$$= (\frac{1}{2} \times 2x^{\frac{1}{2}} \times e^{\frac{1}{2}x}) \times \frac{e^{\frac{1}{2}x}}{x^{\frac{1}{2}}}$$

240

$$= x^{\frac{1}{2}}e^{\frac{1}{2}x} + \frac{e^{\frac{1}{2}x}}{x^{\frac{1}{2}}}$$

$$= \frac{x\,e^{\frac{1}{2}x} + e^{\frac{1}{2}x}}{x^{\frac{1}{2}}}$$

$$= \frac{e^{\frac{1}{2}x}(x+1)}{x^{\frac{1}{2}}}$$

$$= \frac{e^{\frac{x}{2}}(x+1)}{x^{\frac{1}{2}}}$$

$$\frac{dy}{dx} = \frac{\sqrt{e^x}(x+1)}{\sqrt{x}}$$

10. Find the derivative of $2x\log_{10}x$

Solution

$y = 2x\log_{10}x$ (Note that this is similar to $2x\log_a x$)

$u = 2x$

$v = \log_{10}x$

$\dfrac{du}{dx} = 2$

$\dfrac{dv}{dx} = \dfrac{1}{x}\log_{10}e$ (Recall that the derivative of $\log_a x$ is $\dfrac{1}{x}\log_a e$)

$\dfrac{dy}{dx} = u\dfrac{dv}{dx} + v\dfrac{du}{dx}$

$\qquad = 2x\left(\dfrac{1}{x}\log_{10}e\right) + \log_{10}x(2)$

$\qquad = 2\log_{10}e + 2\log_{10}x$

$\dfrac{dy}{dx} = 2(\log_{10}e + \log_{10}x)$

Or, $\dfrac{dy}{dx} = 2\left(\dfrac{\log_e e}{\log_e 10} + \log_{10}x\right)$

 Note that the rule of change of base in logarithm that has been applied above is given by:

If $\log_a b$ is to be converted to a new base, e, then: $\log_a b = \dfrac{\log_e b}{\log_e a}$

$\dfrac{dy}{dx} = 2\left(\dfrac{1}{\log_e 10} + \log_{10}x\right)$ (Recall that $\log_x x = 1$, or $\log_e e = 1$)

11. Find the derivative of 10^{2x}

Solution

$y = 10^{2x}$ (This is similar to a^{2x})

$\dfrac{dy}{dx} = \dfrac{d(2x)}{dx} \times 10^{2x}\log_e 10$

$$= 2 \times 10^{2x} \log_e 10$$

$$= (10^{2x}) 2 \log_e 10$$

$$= 10^{2x} \log_e 10^2 \quad \text{(Recall that } a\log_x y = \log_x y^a)$$

$$\frac{dy}{dx} = 10^{2x} \log_e 100$$

12. If $y = e^{\ln x}$ find $\frac{dy}{dx}$

<u>Solution</u>

$$y = e^{\ln x} \quad \text{(Or } y = e^{\log_e x} \text{ since } \ln x = \log_e x)$$

Let, $u = \ln x$

$$y = e^u$$

$$\frac{du}{dx} = \frac{1}{x}$$

$$\frac{dy}{du} = e^u$$

$$\frac{dy}{dx} = e^u \times \frac{1}{x}$$

$$= \frac{e^u}{x}$$

$$= \frac{e^{\ln x}}{x} \quad \text{(Since } u = \ln x)$$

$$= \frac{e^{\log_e x}}{x} \quad \text{(Since } \ln x = \log_e x)$$

$$= \frac{x}{x} \quad \text{(Recall the identity that } m^{\log_m n} = n)$$

$$\frac{dy}{dx} = 1$$

This example shows that calculus works, since the derivative of x is 1, and the derivative of $e^{\log_e x} = 1$ because $e^{\log_e x} = x$. Hence, we could have solved this example by stating that:

$$\frac{d(e^{\ln x})}{dx} = \frac{d(x)}{dx} = 1$$

13. What is the derivative of $\dfrac{e^{\frac{1}{x}}}{x^2}$

<u>Solution</u>

$$y = \frac{e^{\frac{1}{x}}}{x^2}$$

From quotient rule:

$$u = e^{\frac{1}{x}}$$

$$v = x^2$$

$$\frac{du}{dx} = \frac{d\left(\frac{1}{x}\right)}{dx} \times e^{\frac{1}{x}}$$

$$= \frac{d(x^{-1})}{dx} \times e^{\frac{1}{x}}$$

$$= -1x^{-2} \times e^{\frac{1}{x}}$$

$$= \frac{-1}{x^2} \times e^{\frac{1}{x}}$$

$$\frac{du}{dx} = \frac{-e^{\frac{1}{x}}}{x^2}$$

$$\frac{du}{dx} = 2x$$

$$\frac{dy}{dx} = \frac{v\frac{du}{dx} - u\frac{dv}{dx}}{v^2} \qquad \text{(Quotient rule)}$$

$$= \frac{x^2\left(\frac{-e^{\frac{1}{x}}}{x^2}\right) - e^{\frac{1}{x}}(2x)}{(x^2)^2}$$

$$= \frac{-e^{\frac{1}{x}} - 2x\,e^{\frac{1}{x}}}{x^4} \qquad \text{(Note that } x^2 \text{ has cancelled out)}$$

14. Find the derivative of $e^{-5x} + 5e$

Solution

$$y = e^{-5x} + 5e$$

$$\frac{dy}{dx} = \frac{d(-5x)}{dx} \times e^{-5x} + \frac{d(5e)}{dx}$$

$$= -5e^{-5x} + 0 \qquad \text{(Note that 5e is a constant and the derivative of a constant is zero)}$$

$$\frac{dy}{dx} = -5e^{-5x}$$

15. If $y = e^{(2x+3)^2}$ find $\frac{dy}{dx}$.

Solution

$$y = e^{(2x+3)^2}$$

Let $u = (2x + 3)^2$

Hence, $y = e^u$

$$\frac{du}{dx} = 2(2x + 3)^{2-1} \times \frac{d(2x)}{dx} \qquad \text{(Using chain rule)}$$

$$= 2(2x + 3) \times 2$$

$$= 4(2x + 3)$$

$$\frac{dy}{du} = e^u$$

$$\frac{dy}{dx} = \frac{dy}{du} \times \frac{du}{dx}$$
$$= e^u \times 4(2x+3)$$
$$= e^{(2x+3)^2} \times 4(2x+3) \qquad \text{(Note that } u = (2x+3)^2\text{)}$$
$$\frac{dy}{dx} = 4(2x+3)e^{(2x+3)^2}$$

16. Differentiate with respect to θ, $5^{2\theta}$

Solution

$$y = 5^{2\theta} \qquad \text{(This is like } y = a^{2\theta}\text{)}$$
$$\frac{dy}{d\theta} = \frac{d(2\theta)}{d\theta} \times 5^{2\theta}\log_e 5$$
$$= 2 \times 5^{2\theta}\log_e 5$$
$$= (5^{2\theta})2\log_e 5$$
$$= 5^{2\theta}\log_e 5^2 \qquad \text{(Note that } 2\log_e 5 = \log_e 5^2\text{)}$$
$$\frac{dy}{d\theta} = 5^{2\theta}\log_e 25$$

Exercise 20

1. Differentiate with respect to x: $5a^{2x}$

2. Find the derivative of $a^{5x^3 - x}$

3. If $y = x^4 a^{3x}$, find $\frac{dy}{dx}$.

4. Find the derivative of $\frac{3 + e^{-x}}{e^x}$

5. If $y = 2e^{\sqrt{3x}}$ find $\frac{dy}{dx}$.

6. Find $\frac{dy}{dx}$ if $y = 6x^3 e^{x^2}$

7. Find the derivative of $a^{-5x} + a^{5x}$

8. Find the derivative of $3e^{4x}\log_e 2x^2$

9. Differentiate with respect to x: $3x^2\sqrt{e^{5x}}$

10. Find the derivative of $x^4\log_{10} 7x$

11. Find the derivative of 6^{3x}

12. Find the derivative of $e^{x^2 - 3x}$

13. What is the derivative of $\frac{2e^{\frac{3}{x}}}{10x^5}$

14. Find the derivative of $1 - 2e^{-3x}$

15. If $y = e^{(x^2+x)^3}$ find $\dfrac{dy}{dx}$.

16. Differentiate with respect to x, 2^{x^2}

17. Find the derivative of $\dfrac{a^x - a^{-x}}{a^{-x}}$

18. Find the derivative of $e^x \ln 10x^3$

19. Differentiate with respect to x: $e^{2x}\sqrt{5x}$

20. Find the derivative of $2x^5 \log_{10} 5x^4$

CHAPTER 21
LOGARITHMIC DIFFERENTIATION

A function of x could be raised to an exponent which is also a function of x. An example is x^x. The differentiation of such a function is called logarithmic differentiation.

This type of function is differentiated by first taking the logarithm of both sides of the equation, and then differentiating implicitly before making $\frac{dy}{dx}$ the subject of the formula (i.e. solving for $\frac{dy}{dx}$). Finally, y (i.e. the function) is substituted into the final answer.

Examples

1. Find the derivative of x^x.

Solution

$$y = x^x$$

This is a function of x raised to an exponent which is also a function of x.

Hence, take the logarithm to base e of both sides as follows:

$$\log_e y = \log_e x^x$$

$$\log_e y = x\log_e x \quad \text{(Note that } \log_m n^b = b\log_m n)$$

Now differentiate y implicitly and differentiate the right hand side using product rule. This gives:

$$\frac{1}{y}\frac{dy}{dx} = x\frac{1}{x} + \log_e x(1) \quad \text{(Use of product rule on the right hand side, with u = } x \text{ and v = } \log_e x)$$

$$\frac{1}{y}\frac{dy}{dx} = 1 + \log_e x$$

Multiply both sides by y to make $\frac{dy}{dx}$ the subject of the formula as follows:

$$\frac{dy}{dx} = y(1 + \log_e x)$$

$$\frac{dy}{dx} = x^x(1 + \log_e x) \quad \text{(Since y = } x^x)$$

2. Find the derivative of $x^{\ln x}$.

Solution

$$y = x^{\ln x}$$

Taking the logarithm of both sides gives:

$$\log_e y = \log_e x^{\ln x}$$

$$\log_e y = \ln x\log_e x$$

Differentiate y implicitly and treat the right hand side using product rule as follows:

$$\frac{1}{y}\frac{dy}{dx} = \ln x\left(\frac{1}{x}\right) + \log_e x\left(\frac{1}{x}\right)$$

$$\frac{1}{y}\frac{dy}{dx} = \left(\frac{\ln x}{x}\right) + \left(\frac{\ln x}{x}\right) \qquad \text{(Note that } \log_e x \text{ can also be written as } \ln x)$$

$$\frac{1}{y}\frac{dy}{dx} = \frac{2\ln x}{x}$$

Multiply both sides by y. This gives:

$$\frac{dy}{dx} = y\left(\frac{2\ln x}{x}\right)$$

$$\frac{dy}{dx} = x^{\ln x}\left(\frac{2\ln x}{x}\right) \qquad \text{(Since } y = x^{\ln x})$$

Or, $\dfrac{dy}{dx} = \dfrac{2}{x}x^{\ln x}\ln x$

3. If $y = \ln(2 + x)^x$ find $\dfrac{dy}{dx}$.

Solution

$y = \ln(2 + x)^x$

In this case we are not going to take the logarithm of both sides since there is already logarithm on the right hand side. Hence, we proceed as follows:

$y = \ln(2 + x)^x$

$y = x\log_e(2 + x)$ [Note that $\ln(2 + x)$ can also be written as $\log_e(2 + x)$]

Using product rule gives:

$$\frac{dy}{dx} = x\left(\frac{1}{2 + x}\right) + \log_e(2 + x)(1)$$

$$\frac{dy}{dx} = \frac{x}{2 + x} + \log_e(2 + x)$$

4. Find $\dfrac{dy}{dx}$ if $y = (x^2 - 3)^{5x}$

Solution

$y = (x^2 - 3)^{5x}$

Taking the natural logarithm of both sides gives:

$\log_e y = \log_e(x^2 - 3)^{5x}$

$\log_e y = 5x\log_e(x^2 - 3)$

Using implicit differentiation for the left hand side and product rule for the right hand side gives:

$$\frac{1}{y}\frac{dy}{dx} = 5x\left(\frac{2x}{x^2 - 3}\right) + \log_e(x^2 - 3)(5)$$

$$\frac{1}{y}\frac{dy}{dx} = \frac{10x^2}{x^2 - 3} + 5\log_e(x^2 - 3)$$

Multiply both sides by y to obtain:

$$\frac{dy}{dx} = y\left[\frac{10x^2}{x^2 - 3} + 5\log_e(x^2 - 3)\right]$$

$$\frac{dy}{dx} = (x^2 - 3)^{5x}\left[\frac{10x^2}{x^2 - 3} + 5\log_e(x^2 - 3)\right] \quad [\text{Note that } y = (x^2 - 3)^{5x}]$$

5. If $y = \dfrac{(x^2 - 5)(3x-1)^2}{x^7(2x^3-3)}$

Solution

This is not a function of x raised to an exponent which is also a function of x. This function can be differentiated by the use of product and quotient rule. However, the use of these rules will be a nightmare. Hence, we have to apply logarithmic differentiation. This is done as follows:

$$y = \frac{(x^2 - 5)(3x-1)^2}{x^7(2x^3-3)}$$

Taking the logarithm of both sides gives:

$$\log_e y = \log_e\left[\frac{(x^2 - 5)(3x-1)^2}{x^7(2x^3-3)}\right]$$

In order to continue, we have to recall the theory of logarithm as follows:

1. $\log_b(cd) = \log_b c + \log_b d$

2. $\log_b\left(\dfrac{c}{d}\right) = \log_b c - \log_b d$

3. $\log_b c^d = d\log_b c$

Therefore, we continue the logarithmic differentiation as follows:

$$\log_e y = \log_e\left[\frac{(x^2 - 5)(3x-1)^2}{x^7(2x^3-3)}\right]$$

Applying the theory above gives:

$$\log_e y = \log_e[(x^2 - 5)(3x - 1)^2] - \log_e[x^7(2x^3 - 3)]$$
$$\log_e y = \log_e(x^2 - 5) + \log_e(3x - 1)^2 - [\log_e x^7 + \log_e(2x^3 - 3)]$$
$$\log_e y = \log_e(x^2 - 5) + 2\log_e(3x - 1) - 7\log_e x - \log_e(2x^3 - 3)$$

Differentiating each term accordingly gives:

$$\frac{1}{y}\frac{dy}{dx} = \frac{2x}{x^2 - 5} + \frac{2 \times 3}{3x - 1} - \frac{7}{x} - \frac{6x}{2x^3 - 3}$$

Note that each function of x is differentiated first, and the value obtained is then divided by the original function. Hence, the expression above simplifies to gives:

$$\frac{1}{y}\frac{dy}{dx} = \frac{2x}{x^2 - 5} + \frac{6}{3x - 1} - \frac{7}{x} - \frac{6x}{2x^3 - 3}$$

Multiply both sides by y to obtain $\dfrac{dy}{dx}$ as follows:

$$\frac{dy}{dx} = y\left(\frac{2x}{x^2 - 5} + \frac{6}{3x - 1} - \frac{7}{x} - \frac{6x}{2x^3 - 3}\right)$$

Now replace y with its original value from the question as follows:

$$\frac{dy}{dx} = \left[\frac{(x^2 - 5)(3x-1)^2}{x^7(2x^3-3)}\right]\left(\frac{2x}{x^2 - 5} + \frac{6}{3x - 1} - \frac{7}{x} - \frac{6x}{2x^3 - 3}\right)$$

6. If $y = \dfrac{(2x-1)(x+4)(6x-5)}{(3-x)^2}$

Solution

$$y = \dfrac{(2x-1)(x+4)(6x-5)}{(3-x)^2}$$

This is a complex function where product rule and quotient rule would be difficult to apply.

Hence, we apply logarithmic differentiation as follows:

$$\log_e y = \log_e\left[\dfrac{(2x-1)(x+4)(6x-5)}{(3-x)^2}\right]$$

Applying the theory of logarithm gives:

$$\log_e y = \log_e[(2x-1)(x+4)(6x-5)] - \log_e(3-x)^2$$
$$\log_e y = \log_e(2x-1) + \log_e(x+4) + \log_e(6x-5) - 2\log_e(3-x)$$

Differentiating accordingly gives:

$$\dfrac{1}{y}\dfrac{dy}{dx} = \dfrac{2}{2x-1} + \dfrac{1}{x+4} + \dfrac{6}{6x-5} - \dfrac{2(-1)}{3-x}$$
$$\dfrac{1}{y}\dfrac{dy}{dx} = \dfrac{2}{2x-1} + \dfrac{1}{x+4} + \dfrac{6}{6x-5} + \dfrac{2}{3-x}$$

Multiply both sides by y. This gives:

$$\dfrac{dy}{dx} = y\left(\dfrac{2}{2x-1} + \dfrac{1}{x+4} + \dfrac{6}{6x-5} + \dfrac{2}{3-x}\right)$$

Finally replace y with its original expression as follows:

$$\dfrac{dy}{dx} = \left[\dfrac{(2x-1)(x+4)(6x-5)}{(3-x)^2}\right]\left(\dfrac{2}{2x-1} + \dfrac{1}{x+4} + \dfrac{6}{6x-5} + \dfrac{2}{3-x}\right)$$

7. Find the derivative of x^{x^2}

Solution

$$y = x^{x^2}$$
$$\log_e y = \log_e x^{x^2}$$
$$\log_e y = x^2\log_e x$$
$$\dfrac{1}{y}\dfrac{dy}{dx} = x^2\left(\dfrac{1}{x}\right) + \log_e x(2x)$$
$$\dfrac{1}{y}\dfrac{dy}{dx} = x + 2x\log_e x$$
$$\dfrac{1}{y}\dfrac{dy}{dx} = x(1 + 2\log_e x)$$
$$\dfrac{dy}{dx} = yx(1 + 2\log_e x)$$
$$= x^{x^2}x(1 + 2\log_e x) \qquad (\text{y has been replaced with } x^{x^2})$$
$$\dfrac{dy}{dx} = x^{x^2+1}(1 + 2\log_e x)$$

Note that the exponents of x^{x^2} and x were added to obtain x^{x^2+1} since x is also x^1

249

8. If $y = e^{e^x}$ find $\dfrac{dy}{dx}$.

Solution

$y = e^{e^x}$

$\log_e y = \log_e e^{e^x}$

$\log_e y = e^x \log_e e$

$\log_e y = e^x$ (Recall that logarithm of a number to the same base is 1. Hence, $\log_e e = 1$)

$\dfrac{1}{y}\dfrac{dy}{dx} = e^x$ (Note that the derivative of e^x is e^x)

$\dfrac{dy}{dx} = y e^x$

$\qquad = e^{e^x} e^x$

$\dfrac{dy}{dx} = e^{e^x + x}$ (Their exponents have been added)

9. Find the derivative of 5^{2^x}

Solution

$y = 5^{2^x}$

$\log_e y = \log_e 5^{2^x}$

$\log_e y = 2^x \log_e 5$

Differentiate the left hand side implicitly. On the right hand side, differentiate 2^x and take $\log_e 5$ as a constant. This gives:

$\dfrac{1}{y}\dfrac{dy}{dx} = 2^x \log_e 2 \log_e 5$

Note that 2^x is treated in a similar way as a^x, and recall that the derivative of a^x is $a^x \log_e a$. Hence the derivative of 2^x is $2^x \log_e 2$, and $\log_e 5$ multiplies it since it is a constant. Therefore:

$\dfrac{dy}{dx} = y(2^x \log_e 2 \log_e 5)$

$\qquad = 5^{2^x}(2^x \log_e 2 \log_e 5)$

$\dfrac{dy}{dx} = 5^{2^x}(2^x \log_e 2)(\log_e 5)$

10. If $y = x^{\ln(2x^2 - 5)}$, find $\dfrac{dy}{dx}$.

Solution

$y = x^{\ln(2x^2 - 5)}$

$\log_e y = \log_e x^{\ln(2x^2 - 5)}$

$\log_e y = \ln(2x^2 - 5)\log_e x$

$\dfrac{1}{y}\dfrac{dy}{dx} = \ln(2x^2 - 5)\left(\dfrac{1}{x}\right) + \log_e x\left(\dfrac{4x}{2x^2 - 5}\right)$ (Use of product rule)

$$\frac{1}{y}\frac{dy}{dx} = \frac{1}{x}\ln(2x^2 - 5) + \left(\frac{4x}{2x^2-5}\right)\log_e x$$

$$\frac{dy}{dx} = y\left[\frac{1}{x}\ln(2x^2 - 5) + \left(\frac{4x}{2x^2-5}\right)\log_e x\right]$$

$$\frac{dy}{dx} = x^{\ln(2x^2-5)}\left[\frac{\ln(2x^2-5)}{x} + \left(\frac{4x}{2x^2-5}\right)\log_e x\right]$$

Exercise 21

1. Find the derivative of x^{2x}

2. Find the derivative of $3x^{\ln 2x}$.

3. If $y = \ln(1 - 4x^2)^x$ find $\frac{dy}{dx}$.

4. Find $\frac{dy}{dx}$ if $y = (6x^2 - 5)^{3x}$

5. If $y = \frac{(2x^3 + 1)(x - 2)^3}{3x^2(x^3 - 1)^2}$

6. If $y = \frac{(2x - 1)(x^2 - 2)}{(1-x)(x - 3)^2}$

7. Find the derivative of $5x^{3x^4}$

8. If $y = e^{e^{3x}}$ find $\frac{dy}{dx}$.

9. Find the derivative of 2^{e^x}

10. If $y = x^{\ln(x^2 - 4x)}$, find $\frac{dy}{dx}$.

11. If $y = \ln(10 + 2x^2)^x$ find $\frac{dy}{dx}$.

12. Find $\frac{dy}{dx}$ if $y = (3x^3 - 8)^{2x}$

13. If $y = \frac{(x + 3)^2(x - 3)^2}{(x^3 - 1)}$

14. If $y = 5x^{(\ln x)^2}$, find $\frac{dy}{dx}$.

15. Find the derivative of $10x^{3x^2}$

CHAPTER 22
DERIVATIVE OF ONE FUNCTION WITH RESPECT TO ANOTHER

We can differentiate one function with respect to another as illustrated by the examples shown below.

Examples

1. Differentiate x^{12} with respect to x^7.

<u>Solution</u>

Let $u = x^{12}$

And $v = x^7$

$$\frac{du}{dx} = 12x^{11}$$

$$\frac{dv}{dx} = 7x^6$$

Differentiating x^{12} with respect to x^7 means differentiating u (i.e. x^{12}) with respect to v (i.e. x^7).

This means $\frac{du}{dv}$. This follows the parametric equation rule given by:

$$\frac{du}{dv} = \frac{\frac{du}{dx}}{\frac{dv}{dx}}$$

$$= \frac{12x^{11}}{7x^6}$$

$$= \frac{12}{7}x^5$$

2. Differentiate $2e^x$ with respect to $\ln 2x$.

<u>Solution</u>

The differentiation of $2e^x$ with respect to $\ln 2x$ simply means:

$$\frac{\text{The derivative of } 2e^x}{\text{The derivative of } \ln 2x}$$

Hence, $\dfrac{d(2e^x)}{dx} = 2e^x$

And, $\dfrac{d(\ln 2x)}{dx} = \dfrac{2}{2x} = \dfrac{1}{x}$

Therefore, the differentiation of $2e^x$ with respect to $\ln 2x$ is given by:

$$\frac{2e^x}{\frac{1}{x}}$$

$$= 2xe^x$$

3. Differntiate $\sin 5x$ with respect to $\cos x$.

Solution

Hence, $\dfrac{d(\sin 5x)}{dx} = 5\cos 5x$

And, $\dfrac{d(\cos x)}{dx} = -\sin x$

Therefore, the differentiation of $\sin 5x$ with respect to $\cos x$ is given by:

$$\dfrac{\text{The derivative of } \sin 5x}{\text{The derivative of } \cos x}$$

$$= \dfrac{5\cos 5x}{-\sin x}$$

$$= -\dfrac{5\cos 5x}{\sin x}$$

4. Find the derivative of $2x^2 - 5$ with respect to $4x - 1$

Solution

This is obtained as follows:

$$\dfrac{\dfrac{d(2x^2 - 5)}{dx}}{\dfrac{d(4x - 1)}{dx}}$$

$$= \dfrac{4x}{4}$$

$$= x$$

Exercise 22

1. Differentiate x^3 with respect to x^5.

2. Differentiate e^{5x} with respect to $\ln x^2$

3. Differentiate $\cos x^2$ with respect to $\sin x$.

4. Find the derivative of $4x^3 - 5x^2$ with respect to $(x - 1)^2$

5. Differentiate $7x^4$ with respect to $3x^2$.

6. Differentiate $2a^x$ with respect to e^x.

7. Differentiate $\tan^2 x$ with respect to $\cos 5x$.

8. Find the derivative of $\ln(x - 1)$ with respect to e^{x-1}

9. Differentiate $\log_e x^5$ with respect to e^{5x}.

10. Differentiate $\ln(e^x - e^{-x})$ with respect to $\ln x^x$

CHAPTER 23
HIGHER DERIVATIVES (SUCCESSIVE DIFFERENTIATION)

If f(x) is differentiated it gives the first derivative denoted by f'(x) or $\frac{dy}{dx}$. If we differentiate f'(x), it gives the second derivative denoted by f''(x) or $\frac{d^2y}{dx^2}$ (read as, dee two y dee x squared). Other higher derivatives such as $\frac{d^3y}{dx^3}, \frac{d^4y}{dx^4}$ etc can also be obtained depending on the function.

Examples

1. Find the first, second and third derivatives of $2x^5 - 3x^4 + 5x^2 - 6$

<u>Solution</u>

$$y = 2x^5 - 3x^4 + 5x^2 - 6$$

$$\frac{dy}{dx} = 10x^4 - 12x^3 + 10x$$

$\frac{d^2y}{dx^2}$ is obtained by differentiating $\frac{dy}{dx}$, which means to differentiate $10x^4 - 12x^3 + 10x$.

Hence, $\frac{d^2y}{dx^2} = 40x^3 - 36x^2 + 10$

$\frac{d^3y}{dx^3}$ is obtained by differentiating $40x^3 - 36x^2 + 10$ as follows:

$$\frac{d^3y}{dx^3} = 120x^2 - 72x$$

In summary, the first derivative $\left(\frac{dy}{dx}\right)$ is $10x^4 - 12x^3 + 10x$, the second derivative $\left(\frac{d^2y}{dx^2}\right)$ is $40x^3 - 36x^2 + 10$, while the third derivative $\left(\frac{d^3y}{dx^3}\right)$ is $120x^2 - 72x$.

2. If $y = \ln x^2$, find $\frac{d^2y}{dx^2}$.

<u>Solution</u>

$$y = \ln x^2$$

$$\frac{dy}{dx} = \frac{2x}{x^2}$$

$$= \frac{2}{x}$$

$$\frac{d^2y}{dx^2} = \frac{d\left(\frac{2}{x}\right)}{dx}$$

$$= \frac{d(2x^{-1})}{dx}$$

$$= -2x^{-2}$$

$$\frac{d^2y}{dx^2} = \frac{-2}{x^2}$$

3. Find $\dfrac{d^3y}{dx^3}$ given that $y = e^{x^3}$

Solution

$$y = e^{x^3}$$

$$\frac{dy}{dx} = 3x^2 e^{x^3} \qquad \text{(Note that } 3x^2 \text{ is from the derivative of } x^3)$$

We now use product rule to obtain $\dfrac{d^2y}{dx^2}$ as follows:

$$\frac{d^2y}{dx^2} = 3x^2 \left[\frac{d(e^{x^3})}{dx} \right] + e^{x^3} \left[\frac{d(3x^2)}{dx} \right]$$

$$= 3x^2 (3x^2 e^{x^3}) + e^{x^3} (6x)$$

$$= 9x^4 e^{x^3} + 6x e^{x^3}$$

$$\frac{d^2y}{dx^2} = e^{x^3} (9x^4 + 6x)$$

Finally, let us also use product rule to obtain $\dfrac{d^3y}{dx^3}$ as follows:

$$\frac{d^3y}{dx^3} = e^{x^3} \left[\frac{d(9x^4 + 6x)}{dx} \right] + (9x^4 + 6x) \left[\frac{d(e^{x^3})}{dx} \right]$$

$$= e^{x^3} (36x^3 + 6) + (9x^4 + 6x)(3x^2 e^{x^3})$$

$$= 36x^3 e^{x^3} + 6 e^{x^3} + 27x^6 e^{x^3} + 18x^3 e^{x^3}$$

$$= 27x^6 e^{x^3} + 36x^3 e^{x^3} + 18x^3 e^{x^3} + 6 e^{x^3}$$

$$= 27x^6 e^{x^3} + 54x^3 e^{x^3} + 6 e^{x^3}$$

$$\frac{d^3y}{dx^3} = 3 e^{x^3} (9x^6 + 18x^3 + 2)$$

4. If $y = \sin 2x^3$, find the third derivative of y.

Solution

$$y = \sin 2x^3$$

$$\frac{dy}{dx} = \cos 2x^3 \frac{d(2x^3)}{dx} \qquad \text{(Recall that the derivative of } \sin x \text{ is } \cos x)$$

$$= \cos 2x^3 (6x^2)$$

$$\frac{dy}{dx} = 6x^2 \cos 2x^3$$

We now use product rule to obtain $\dfrac{d^2y}{dx^2}$ as follows:

$$\frac{d^2y}{dx^2} = 6x^2 \left[\frac{d(\cos 2x^3)}{dx} \right] + \cos 2x^3 \left[\frac{d(6x^2)}{dx} \right]$$

255

$$= 6x^2\left[-\sin 2x^3 \frac{d(2x^3)}{dx}\right] + \cos 2x^3(12x)$$

$$= 6x^2\left[-\sin 2x^3 (6x^2)\right] + 12x\cos 2x^3$$

$$\frac{d^2y}{dx^2} = -36x^4 \sin 2x^3 + 12x\cos 2x^3$$

Again, the use of product rule gives us $\frac{d^3y}{dx^3}$ as follows:

$$\frac{d^3y}{dx^3} = -36x^4\left[\frac{d(\sin 2x^3)}{dx}\right] + \sin 2x^3\left[\frac{d(-36x^4)}{dx}\right] + 12x\left[\frac{d(\cos 2x^3)}{dx}\right] + \cos 2x^3\left[\frac{d(12x)}{dx}\right]$$

Note that the derivative of $\sin 2x^3$ is $6x^2\cos 2x^3$ as obtained from $\frac{dy}{dx}$. Similarly, the differentiation of $\cos 2x^3$ will give us $-6x^2\sin 2x^3$ (since the derivative of $\sin x$ and $\cos x$ differs only by sign and the interchanging of sin with cos or cos with sin).

We now replace the derivative of $\sin 2x^3$ and $\cos 2x^3$ with $6x^2\cos 2x^3$ and $-6x^2\sin 2x^3$ respectively in the expression above. This gives:

$$\frac{d^3y}{dx^3} = -36x^4(6x^2\cos 2x^3) + \sin 2x^3(-144x^3) + 12x(-6x^2\sin 2x^3) + \cos 2x^3(12)$$

$$= -216x^6\cos 2x^3 - 144x^3\sin 2x^3 - 72x^3\sin 2x^3 + 12\cos 2x^3$$

$$\frac{d^3y}{dx^3} = -216x^6\cos 2x^3 - 216x^3\sin 2x^3 + 12\cos 2x^3$$

Or, $\frac{d^3y}{dx^3} = -12(18x^6\cos 2x^3 + 18x^3\sin 2x^3 - \cos 2x^3)$

5. Find $\frac{d^2y}{dx^2}$ given that $y = a^{x^2} - 5$

<u>Solution</u>

$$y = a^{x^2} - 5$$

$$\frac{dy}{dx} = a^{x^2}\left[\frac{d(x^2)}{dx}\right]\log_e a$$

$$= a^{x^2}(2x)\log_e a$$

$$\frac{dy}{dx} = 2x\, a^{x^2}\log_e a$$

We now use product rule to find $\frac{d^2y}{dx^2}$ as follows:

$$\frac{d^2y}{dx^2} = 2x\left[\frac{d\left(a^{x^2}\log_e a\right)}{dx}\right] + a^{x^2}\log_e a\left[\frac{d(2x)}{dx}\right]$$

$$= 2x\left[\frac{\log_e a\, d\left(a^{x^2}\right)}{dx}\right] + a^{x^2}\log_e a(2)$$

$$= 2x[\log_e a(2xa^{x^2}\log_e a)] + 2a^{x^2}\log_e a$$

$$= 4x^2 a^{x^2}(\log_e a)^2 + 2a^{x^2}\log_e a$$

$$\frac{d^2y}{dx^2} = 2a^{x^2}\log_e a(2x^2\log_e a + 1)$$

6. Find the second derivative of $y = \dfrac{\cos x}{x^3}$

Solution

$$y = \frac{\cos x}{x^3}$$

We use quotient rule as follows:

$$\frac{dy}{dx} = \frac{x^3\frac{d(\cos x)}{dx} - \cos x\frac{d(x^3)}{dx}}{(x^3)^2}$$

$$= \frac{x^3(-\sin x) - \cos x(3x^2)}{x^6}$$

$$\frac{dy}{dx} = \frac{-x^3\sin x - 3x^2\cos x}{x^6}$$

Divide each part by x^6 to separate into fractions as follows:

$$\frac{dy}{dx} = \frac{-x^3\sin x}{x^6} - \frac{3x^2\cos x}{x^6}$$

$$\frac{dy}{dx} = \frac{-\sin x}{x^3} - \frac{3\cos x}{x^4}$$

We now apply quotient rule again as follows:

$$\frac{d^2y}{dx^2} = \frac{x^3\frac{d(-\sin x)}{dx} - (-\sin x)\frac{d(x^3)}{dx}}{(x^3)^2} - \left[\frac{x^4\frac{d(3\cos x)}{dx} - 3\cos x\frac{d(x^4)}{dx}}{(x^4)^2}\right]$$

$$= \frac{x^3(-\cos x) + \sin x(3x^2)}{x^6} - \left[\frac{x^4(-3\sin x) - 3\cos x(4x^3)}{x^8}\right]$$

$$= \frac{-x^3\cos x + 3x^2\sin x}{x^6} + \frac{3x^4\sin x + 12x^3\cos x}{x^8}$$

Separating into fractions gives:

$$\frac{d^2y}{dx^2} = \frac{-x^3\cos x}{x^6} + \frac{3x^2\sin x}{x^6} + \frac{3x^4\sin x}{x^8} + \frac{12x^3\cos x}{x^8}$$

$$= \frac{-\cos x}{x^3} + \frac{3\sin x}{x^4} + \frac{3\sin x}{x^4} + \frac{12\cos x}{x^5}$$

Combining the fractions again by using x^5 as LCM gives:

$$\frac{d^2y}{dx^2} = \frac{-x^2\cos x + 3x\sin x + 3x\sin x + 12\cos x}{x^5}$$

$$\frac{d^2y}{dx^2} = \frac{6x\sin x + 12\cos x - x^2\cos x}{x^5}$$

7. Find the third derivative of $y = \log_e(1 + 2x)^2$

Solution

$$y = \log_e(1 + 2x)^2$$

$$\frac{dy}{dx} = \frac{2(1+2x)^{2-1} \times \frac{d(2x)}{dx}}{(1+2x)^2}$$ [Note the use of chain rule in finding the derivative of $(1 + 2x)^2$]

$$= \frac{2(1+2x) \times 2}{(1+2x)^2}$$

$$= \frac{4(1+2x)}{(1+2x)^2}$$

$$= \frac{4}{1+2x}$$ $(1 + 2x$ cancels out)

$$\frac{dy}{dx} = 4(1 + 2x)^{-1}$$

Using chain rule gives $\frac{d^2y}{dx^2}$ as follows:

$$\frac{d^2y}{dx^2} = -1 \times 4(1 + 2x)^{-1-1} \times \frac{d(2x)}{dx}$$

$$= -4(1 + 2x)^{-2} \times 2$$

$$= -8(1 + 2x)^{-2}$$

Use of chain rule again gives $\frac{d^3y}{dx^3}$ as follows:

$$\frac{d^3y}{dx^3} = -2 \times -8(1 + 2x)^{-2-1} \times \frac{d(2x)}{dx}$$

$$= 16(1 + 2x)^{-3} \times 2$$

$$= 32(1 + 2x)^{-3}$$

$$\frac{d^3y}{dx^3} = \frac{32}{(1+2x)^3}$$

8. If y = sec2x, find $\frac{d^3y}{dx^3}$.

Solution

   y = sec2x

$$\frac{dy}{dx} = 2sec2x tan2x$$

We now use product rule to obtain $\frac{d^2y}{dx^2}$ as follows:

$$\frac{d^2y}{dx^2} = 2sec2x \frac{d(\tan 2x)}{dx} + \tan2x \frac{d(2sec 2x)}{dx}$$

$$= 2sec2x(2sec^2 2x) + \tan2x(2 \times 2\, sec2x tan2x)$$

$$\frac{d^2y}{dx^2} = 4sec^3 2x + 4sec2x tan^2 2x$$

We now use chain rule for $4sec^3 2x$ and product rule for $4sec2x tan^2 2x$ to obtain $\frac{d^3y}{dx^3}$ as follows:

$$\frac{d^3y}{dx^3} = 3 \times 4\, sec^2 2x \frac{d(\sec 2x)}{dx} + 4sec2x \frac{d(\tan^2 2x)}{dx} + \tan^2 2x \frac{d(4sec 2x)}{dx}$$

$$= 12sec^2 2x(2sec2x tan2x) + 4sec2x \left[2tan2x \frac{d(\tan2x)}{dx} \right] + \tan^2 2x[4(2sec2x tan2x)]$$

$$= 24\sec^3 2x\tan 2x + 8\sec 2x\tan 2x(2\sec^2 2x) + 8\sec 2x\tan^3 2x$$

$$= 24\sec^3 2x\tan 2x + 16\sec^3 2x\tan 2x + 8\sec 2x\tan^3 2x$$

$$\frac{d^3y}{dx^3} = 40\sec^3 2x\tan 2x + 8\sec 2x\tan^3 2x$$

9. Given that $y = 2\cos 5x + 3\sin 5x$, prove that $\dfrac{d^2y}{dx^2} + 25y = 0$

Solution

$$y = 2\cos 5x + 3\sin 5x$$

$$\frac{dy}{dx} = 2 \times 5(-\sin 5x) + 3 \times 5(\cos 5x)$$

$$= -10\sin 5x + 15\cos 5x$$

$$\frac{d^2y}{dx^2} = -10 \times 5(\cos 5x) + 15 \times 5(-\sin 5x)$$

$$= -50\cos 5x - 75\sin 5x$$

We now obtain 25y as follows:

$$y = 2\cos 5x + 3\sin 5x$$

$$25y = 25(2\cos 5x + 3\sin 5x)$$

$$25y = 50\cos 5x + 75\sin 5x$$

We now simplify $\dfrac{d^2y}{dx^2} + 25y$ as follows:

$$\frac{d^2y}{dx^2} + 25y = -50\cos 5x - 75\sin 5x + 50\cos 5x + 75\sin 5x$$

$$= 0 \quad \text{(Equal terms with opposite signs cancel out)}$$

Therefore, $\dfrac{d^2y}{dx^2} + 25y = 0$ (As proven above)

10. If $y = x + \sqrt{4 + x^2}$, show that $(4 + x^2)\dfrac{d^2y}{dx^2} + x\dfrac{dy}{dx} - y = 0$

Solution

$$y = x + \sqrt{4 + x^2}$$

$$y = x + (4 + x^2)^{\frac{1}{2}}$$

$$\frac{dy}{dx} = 1 + \frac{1}{2}(4 + x^2)^{\frac{1}{2} - 1} \times \frac{d(x^2)}{dx}$$

$$= 1 + \frac{1}{2}(4 + x^2)^{-\frac{1}{2}} \times 2x$$

$$\frac{dy}{dx} = 1 + x(4 + x^2)^{-\frac{1}{2}}$$

Using product rule, we obtain $\dfrac{d^2y}{dx^2}$ as follows:

$$\frac{d^2y}{dx^2} = 0 + x\left[\frac{d(4+x^2)^{-\frac{1}{2}}}{dx}\right] + (4+x^2)^{-\frac{1}{2}}\frac{d(x)}{dx}$$

$$= x\left[-\frac{1}{2}(4+x^2)^{-\frac{1}{2}-1}\frac{d(x^2)}{dx}\right] + (4+x^2)^{-\frac{1}{2}}\ (1)$$

$$= x\left[-\frac{1}{2}(4+x^2)^{-\frac{3}{2}} \times 2x\right] + (4+x^2)^{-\frac{1}{2}}$$

$$= -x^2(4+x^2)^{-\frac{3}{2}} + (4+x^2)^{-\frac{1}{2}}$$

$$\frac{d^2y}{dx^2} = \frac{-x^2}{(4+x^2)^{\frac{3}{2}}} + \frac{1}{(4+x^2)^{\frac{1}{2}}}$$

Let us now simplify $(4+x^2)\dfrac{d^2y}{dx^2} + x\dfrac{dy}{dx} - y$ as follows:

$$(4+x^2)\left[\frac{-x^2}{(4+x^2)^{\frac{3}{2}}} + \frac{1}{(4+x^2)^{\frac{1}{2}}}\right] + x[1 + x(4+x^2)^{-\frac{1}{2}}] - [x + (4+x^2)^{\frac{1}{2}}]$$

Expanding the brackets gives:

$$-x^2(4+x^2)^{1-\frac{3}{2}} + (4+x^2)^{1-\frac{1}{2}} + x + x^2(4+x^2)^{-\frac{1}{2}} - x - (4+x^2)^{\frac{1}{2}}$$

$$= -x^2(4+x^2)^{-\frac{1}{2}} + (4+x^2)^{\frac{1}{2}} + x + x^2(4+x^2)^{-\frac{1}{2}} - x - (4+x^2)^{\frac{1}{2}}$$

$$= 0 \quad \text{(Note that equal terms with opposite signs cancel out one another to give zero)}$$

Therefore, $(4+x^2)\dfrac{d^2y}{dx^2} + x\dfrac{dy}{dx} - y = 0$ (As proven above)

Exercise 23

1. Find the first, second and third derivatives of $x^4 - 2x^3 - 7x^2 + 1$

2. If $y = \ln 2x^3$, find $\dfrac{d^2y}{dx^2}$.

3. Find $\dfrac{d^2y}{dx^2}$ given that $y = e^{5x^4}$

4. If $y = \sin^2 x^5$, find the second derivative of y.

5. Find $\dfrac{d^3y}{dx^3}$ given that $y = \ln(4x - 10)$

6. Find the second derivative of $y = \dfrac{\sin x}{x}$

7. Find the third derivative of $y = (\ln x)^2$

8. If $y = \cot x$, find $\dfrac{d^3y}{dx^3}$.

9. Given that $y = \sin^2 x + \cos 2x$, find $\dfrac{d^2y}{dx^2} - y$

10. If $y = 2e^{2x} + 5e^{-x}$ evaluate $\dfrac{d^2y}{dx^2} - \dfrac{dy}{dx} - 2y$

11. Given that $y = \cos^2 x + 2\sin x^2$ find $\dfrac{d^2y}{dx^2} - \dfrac{dy}{dx}$

12. Find the second derivative of $y = \dfrac{\tan x}{3x}$

13. If $y = e^{-x}\cos 2x$, find the second derivative of y.

14. If $y = 2\cos 3x + 5\sin 3x$, evaluate $\dfrac{d^2y}{dx^2} + 9y$

15. Find the second derivative of $y = \dfrac{\sin 3x}{3x}$

CHAPTER 24
MISCELLANEOUS PROBLEMS ON DIFFERENTIAL CALCULUS

This chapter covers worked examples on general problems involving combination of topics treated in the various previous chapters. More challenging problems would also be covered here.

Examples

1. If $y = \dfrac{1}{(2x^3 - 5)^4}$ find $\dfrac{dy}{dx}$.

Solution

$$y = \frac{1}{(2x^3 - 5)^4}$$

This can be written as:

$$y = (2x^3 - 5)^{-4}$$

We now use chain rule to differentiate it as follows:

$$\frac{dy}{dx} = -4(2x^3 - 5)^{-4-1} \times \frac{d(2x^3 - 5)}{dx}$$

$$= -4(2x^3 - 5)^{-5} \times 6x^2$$

$$= -24x^2(2x^3 - 5)^{-5}$$

$$\frac{dy}{dx} = \frac{-24x^2}{(2x^3 - 5)^5}$$

2. Given that $f(x) = 5x^4 - 3x^3 - 7x^2 + 9$, find $f'(2)$

Solution

$$f(x) = 5x^4 - 3x^3 - 7x^2 + 9$$

We differentiate $f(x)$ to obtain $f'(x)$ as follows:

$$f'(x) = 20x^3 - 9x^2 - 14x$$

In order to find $f'(2)$, we simply substitute 2 for x in $f'(x)$ as follows:

$$f'(x) = 20x^3 - 9x^2 - 14x$$

$$f'(2) = 20(2^3) - 9(2^2) - 14(2)$$

$$= 20(8) - 9(4) - 28$$

$$= 160 - 36 - 28$$

$$f'(2) = 96$$

3. A function $f(x)$ is given by $f(x) = \dfrac{\sqrt{4 + 3x^2}}{x^3}$. Find:

(a) the derivative of $f(x)$

(b) the gradient of $f(x)$ at the point $(2, \frac{1}{2})$

Solution

(a) $f(x) = \dfrac{\sqrt{4 + 3x^2}}{x^3}$

We apply quotient rule to differentiate f(x) as follows:

$$f'(x) = \frac{x^3\left[\dfrac{d(\sqrt{4 + 3x^2})}{dx}\right] - \sqrt{4 + 3x^2}\left[\dfrac{d(x^3)}{dx}\right]}{(x^3)^2}$$

$$= \frac{x^3\left[\dfrac{d(4 + 3x^2)^{\frac{1}{2}}}{dx}\right] - (4 + 3x^2)^{\frac{1}{2}}(3x^2)}{x^6}$$

$$= \frac{x^3\left[\dfrac{1}{2}(4 + 3x^2)^{\frac{1}{2} - 1} \times \dfrac{d(3x^2)}{dx}\right] - 3x^2(4 + 3x^2)^{\frac{1}{2}}}{x^6}$$

$$= \frac{x^3\left[\dfrac{1}{2}(4 + 3x^2)^{-\frac{1}{2}} \times 6x\right] - 3x^2(4 + 3x^2)^{\frac{1}{2}}}{x^6}$$

$$= \frac{x^3\left[3x(4 + 3x^2)^{-\frac{1}{2}}\right] - 3x^2(4 + 3x^2)^{\frac{1}{2}}}{x^6}$$

$$= \frac{3x^4(4 + 3x^2)^{-\frac{1}{2}} - 3x^2(4 + 3x^2)^{\frac{1}{2}}}{x^6}$$

Let us factorize the expression above by taking $3x^2(4 + 3x^2)^{-\frac{1}{2}}$ as the common factor. This gives:

$$f'(x) = \frac{3x^2(4 + 3x^2)^{-\frac{1}{2}}[x^2 - (4 + 3x^2)]}{x^6}$$

Note that $(4 + 3x^2)^{\frac{1}{2}} \div (4 + 3x^2)^{-\frac{1}{2}} = (4 + 3x^2)^{\frac{1}{2} - (-\frac{1}{2})} = 4 + 3x^2$ as the exponent becomes 1. Recall that exponents are subtracted during division.

$$= \frac{3x^2(4 + 3x^2)^{-\frac{1}{2}}(x^2 - 4 - 3x^2)}{x^6}$$

$$= \frac{3x^2(4 + 3x^2)^{-\frac{1}{2}}(-2x^2 - 4)}{x^6}$$

$$= \frac{-3x^2(4 + 3x^2)^{-\frac{1}{2}}(2x^2 + 4)}{x^6}$$

$$= \frac{-3(4 + 3x^2)^{-\frac{1}{2}}(2x^2 + 4)}{x^4} \qquad \text{(Note that } x^2 \text{ has cancelled out of } x^6\text{)}$$

$$f'(x) = \frac{-3(2x^2 + 4)}{x^4(4 + 3x^2)^{\frac{1}{2}}}$$

(b) The gradient of f(x) at the point $(2, \frac{1}{2})$ is obtained by substituting 2 for x in f'(x). Note that

the point $(2, \frac{1}{2})$ is at $x = 2$ and $y = \frac{1}{2}$. We ignore $y = \frac{1}{2}$ since y is not in the expression for f'(x).

Hence, $f'(x) = \dfrac{-3(2x^2 + 4)}{x^4(4 + 3x^2)^{\frac{1}{2}}}$

At $(2, \frac{1}{2})$, $f'(2) = \dfrac{-3(2(2)^2 + 4)}{(2)^4(4 + 3(2)^2)^{\frac{1}{2}}}$

$= \dfrac{-3(8 + 4)}{16(4 + 12)^{\frac{1}{2}}}$

$= \dfrac{-3(12)}{16(16)^{\frac{1}{2}}}$

$= \dfrac{-36}{16 \times 4}$ (Note that $(16)^{\frac{1}{2}} = \sqrt{16} = 4$)

$f'(2) = \dfrac{-9}{16}$ (In its lowest term)

Therefore the gradient of f(x) at the point $(2, \frac{1}{2})$ is $\dfrac{-9}{16}$

4. If $x^2y^2 - 3xy + 4xy^3 = 4$, find:

(a) the derivative of the expression

(b) the gradient at $(-1, 2)$

Solution

(a) $x^2y^2 - 3xy + 4xy^3 = 4$

The use of implicit differentiation combined with product rule gives us the derivative as follows:

$x^2\dfrac{d(y^2)}{dx} + y^2\dfrac{d(x^2)}{dx} - \left[3x\dfrac{d(y)}{dx} + y\dfrac{d(3x)}{dx}\right] + 4x\dfrac{d(y^3)}{dx} + y^3(4) = \dfrac{d(4)}{dx}$

$x^2\left(2y\dfrac{dy}{dx}\right) + y^2(2x) - \left[3x\dfrac{dy}{dx} + y(3)\right] + 4x\left(3y^2\dfrac{dy}{dx}\right) + 4y^3 = 0$

$2x^2y\dfrac{dy}{dx} + 2xy^2 - 3x\dfrac{dy}{dx} - 3y + 12xy^2\dfrac{dy}{dx} + 4y^3 = 0$

Collect terms in $\dfrac{dy}{dx}$ on one side of the equation. This gives:

$2x^2y\dfrac{dy}{dx} - 3x\dfrac{dy}{dx} + 12xy^2\dfrac{dy}{dx} = 3y - 2xy^2 - 4y^3$

Factorizing the left hand side gives:

$\dfrac{dy}{dx}(2x^2y - 3x + 12xy^2) = 3y - 2xy^2 - 4y^3$

Divide both sides by the terms in the bracket. This gives:

$\dfrac{dy}{dx} = \dfrac{3y - 2xy^2 - 4y^3}{12x^2y - 3x + 12xy^2}$

(b) At the point $(-1, 2)$ the gradient of the expression is obtained by simply substituting -1 for x and 2 for y in the expression for the derivative. This is done as follows:

$$\frac{dy}{dx} = \frac{3y - 2xy^2 - 4y^3}{12x^2y - 3x + 12xy^2}$$

$$= \frac{3(2) - 2(-1)(2)^2 - 4(2)^3}{12(-1)^2(2) - 3(-1) + 12(-1)(2)^2}$$

$$= \frac{6 + 8 - 32}{24 + 3 - 48}$$

$$= \frac{-18}{-21}$$

$$= \frac{6}{7}$$

5. Find the derivative of $(x + 3y^2)^3 = 7$

Solution

$$(x + 3y^2)^3 = 7$$

We differentiate implicitly and apply chain rule as follows:

$$3(x + 3y^2)^{3-1} \times \frac{d(x+3y^2)}{dx} = 0$$

$$3(x + 3y^2)^2\left(1 + 6y\frac{dy}{dx}\right) = 0$$

Dividing both sides by $3(x + 3y^2)^2$ gives:

$$1 + 6y\frac{dy}{dx} = 0 \qquad \text{(Note that } \frac{0}{3(x + 3y^2)^2} = 0)$$

$$6y\frac{dy}{dx} = -1$$

$$\frac{dy}{dx} = -\frac{1}{6y}$$

6. If $y = (3x^2 - 2x + 5)(2x - 3)$, find $\frac{dy}{dx}$.

Solution

$$y = (3x^2 - 2x + 5)(2x - 3)$$

We differentiate the expression by applying product rule as follows:

$$\frac{dy}{dx} = (3x^2 - 2x + 5)(2) + (2x - 3)(6x - 2)$$

$$= 6x^2 - 4x + 10 + 12x^2 - 4x - 18x + 6$$

$$\frac{dy}{dx} = 18x^2 - 26x + 16$$

7. Find the derivative of $a^{1 + \tan x}$

Solution

$$y = a^{1 + \tan x}$$

266

Let u = 1 + tan$x$

Hence, y = a^u

$$\frac{du}{dx} = \sec^2 x$$

$$\frac{dy}{du} = a^u \log_e a$$

Hence, $\dfrac{dy}{dx} = \dfrac{dy}{du} \times \dfrac{du}{dx}$

$$= a^u \log_e a \times \sec^2 x$$

$$\frac{dy}{dx} = a^{1+\tan x} \sec^2 x \log_e a$$

8. Find the derivative of $\dfrac{\log_e x}{1+\cos x}$

Solution

$$y = \frac{\log_e x}{1+\cos x}$$

We apply product rule as follows:

$$\frac{dy}{dx} = \frac{(1+\cos x)\frac{d(\log_e x)}{dx} - \log_e x \frac{d(1+\cos x)}{dx}}{(1+\cos x)^2}$$

$$= \frac{(1+\cos x)\frac{1}{x} - \log_e x(-\sin x)}{(1+\cos x)^2}$$

$$= \frac{\frac{1+\cos x}{x} + \sin x \log_e x}{(1+\cos x)^2}$$

$$= \frac{\frac{1+\cos x + x \sin x \log_e x}{x}}{(1+\cos x)^2}$$

$$\frac{dy}{dx} = \frac{1+\cos x + x \sin x \log_e x}{x(1+\cos x)^2}$$

9. Differentiate with respect to x: ln(cos$x$ + sin$x$)

Solution

$$y = \ln(\cos x + \sin x)$$

$$\frac{dy}{dx} = \frac{\frac{d(\cos x + \sin x)}{dx}}{\cos x + \sin x}$$

$$= \frac{-\sin x + \cos x}{\cos x + \sin x}$$

$$\frac{dy}{dx} = \frac{\cos x - \sin x}{\cos x + \sin x}$$

10. If $y = e^{\sin 2x + \cos x}$ find $\frac{dy}{dx}$.

Solution

$$y = e^{\sin 2x + \cos x}$$

$$\frac{dy}{dx} = \frac{d(\sin 2x + \cos x)}{dx} \times e^{\sin 2x + \cos x}$$

$$= (2\cos 2x - \sin x)(e^{\sin 2x + \cos x})$$

11. Given that $y = e^x - e^{-x}$ show that $\frac{d^3y}{dx^3} + \frac{d^2y}{dx^2} + \frac{dy}{dx} + y = 4e^x$

Solution

$$y = e^x - e^{-x}$$

$$\frac{dy}{dx} = e^x - (-e^{-x}) \qquad \left(\text{Note that } \frac{d(e^{-x})}{dx} = \frac{d(-x)}{dx} \times e^{-x} = -1 \times e^{-x} = -e^{-x}\right)$$

$$= e^x + e^{-x}$$

$$\frac{d^2y}{dx^2} = e^x - e^{-x}$$

$$\frac{d^3y}{dx^3} = e^x + e^{-x}$$

Let us now substitute corresponding term into $\frac{d^3y}{dx^3} + \frac{d^2y}{dx^2} + \frac{dy}{dx} + y$ as follows:

$$(e^x + e^{-x}) + (e^x - e^{-x}) + (e^x + e^{-x}) + (e^x - e^{-x})$$

Collecting like terms together gives:

$$e^x + e^x + e^x + e^x + e^{-x} - e^{-x} + e^{-x} - e^{-x}$$

$$= 4e^x \qquad (\text{Note that } e^{-x} \text{ cancels out each other})$$

Hence, $\frac{d^3y}{dx^3} + \frac{d^2y}{dx^2} + \frac{dy}{dx} + y = 4e^x \qquad (\text{As proven above})$

12. Differentiate with respect to x: $\ln\left(\frac{1 - 3x^2}{1 + 3x^2}\right)^{\frac{1}{2}}$

Solution

$$y = \ln\left(\frac{1 - 3x^2}{1 + 3x^2}\right)^{\frac{1}{2}}$$

$$= \log_e\left(\frac{1 - 3x^2}{1 + 3x^2}\right)^{\frac{1}{2}} \qquad (\text{Note that "ln" is } \log_e)$$

$$= \log_e\left[\frac{(1 - 3x^2)^{\frac{1}{2}}}{(1 + 3x^2)^{\frac{1}{2}}}\right]$$

$$= \log_e(1 - 3x^2)^{\frac{1}{2}} - \log_e(1 + 3x^2)^{\frac{1}{2}}$$

$$y = \frac{1}{2}\log_e(1 - 3x^2) - \frac{1}{2}\log_e(1 + 3x^2)$$

$$\frac{dy}{dx} = \frac{1}{2}\frac{\frac{d(1-3x^2)}{dx}}{1-3x^2} - \frac{1}{2}\frac{\frac{d(1+3x^2)}{dx}}{1+3x^2}$$

$$= \frac{1}{2}\frac{-6x}{1-3x^2} - \frac{1}{2}\frac{6x}{1+3x^2}$$

$$= \frac{-3x}{1-3x^2} - \frac{3x}{1+3x^2} \qquad \left(\text{Note that } \frac{1}{2} \text{ reduces } 6x \text{ to } 3x\right)$$

$$= \frac{-3x(1+3x^2) - 3x(1-3x^2)}{(1-3x^2)(1+3x^2)}$$

$$= \frac{-3x - 9x^3 - 3x + 9x^3)}{(1-3x^2)(1+3x^2)}$$

$$= \frac{-6x}{1 + 3x^2 - 3x^2 - 9x^4}$$

$$\frac{dy}{dx} = \frac{-6x}{1 - 9x^4}$$

13. If $y = \dfrac{x}{\sqrt{9 - x^2}}$ show that: $(9 - x^2)\dfrac{d^2y}{dx^2} = 3x\dfrac{dy}{dx}$

<u>Solution</u>

$$y = \frac{x}{\sqrt{9 - x^2}}$$

Applying quotient rule gives $\dfrac{dy}{dx}$ as follows:

$$\frac{dy}{dx} = \frac{\sqrt{9 - x^2}(1) - x\frac{d(\sqrt{9 - x^2})}{dx}}{(\sqrt{9 - x^2})^2}$$

$$= \frac{(9 - x^2)^{\frac{1}{2}}(1) - x\frac{d[(9 - x^2)^{\frac{1}{2}}]}{dx}}{[(9 - x^2)^{\frac{1}{2}}]^2}$$

$$= \frac{(9 - x^2)^{\frac{1}{2}} - x\left[\frac{1}{2}(9 - x^2)^{\frac{1}{2} - 1}\right] \times (-2x)}{9 - x^2} \qquad (\text{Note that } -2x \text{ is from the derivative of } -x^2)$$

$$= \frac{(9 - x^2)^{\frac{1}{2}} - x\left[-x(9 - x^2)^{-\frac{1}{2}}\right]}{9 - x^2}$$

$$= \frac{(9 - x^2)^{\frac{1}{2}} + x^2(9 - x^2)^{-\frac{1}{2}}}{9 - x^2}$$

$$= \frac{(9 - x^2)^{\frac{1}{2}} + \dfrac{x^2}{(9 - x^2)^{\frac{1}{2}}}}{9 - x^2}$$

$$= \frac{\dfrac{9 - x^2 + x^2}{(9 - x^2)^{\frac{1}{2}}}}{9 - x^2}$$

$$= \frac{9}{(9 - x^2)^{\frac{1}{2}}(9 - x^2)}$$

$$= \frac{9}{(9 - x^2)^{\frac{3}{2}}}$$

$$\frac{dy}{dx} = 9(9 - x^2)^{-\frac{3}{2}}$$

Applying chain rule gives $\dfrac{d^2y}{dx^2}$ as follows:

$$\frac{d^2y}{dx^2} = \frac{-3}{2} \times 9(9 - x^2)^{-\frac{3}{2} - 1} \times -2x \qquad \text{(Note that } -2x \text{ is from the derivative of } -x^2\text{)}$$

$$= \frac{-27}{2} \times -2x(9 - x^2)^{-\frac{5}{2}}$$

$$= 27x(9 - x^2)^{-\frac{5}{2}}$$

$$\frac{d^2y}{dx^2} = \frac{27x}{(9 - x^2)^{\frac{5}{2}}}$$

Let us now show that $(9 - x^2)\dfrac{d^2y}{dx^2} = 3x\dfrac{dy}{dx}$

We simplify $(9 - x^2)\dfrac{d^2y}{dx^2}$ as follows:

$$(9 - x^2)\frac{27x}{(9 - x^2)^{\frac{5}{2}}}$$

$$= \frac{27x(9 - x^2)}{(9 - x^2)^{\frac{5}{2}}}$$

$$= 27x(9 - x^2)^{1 - \frac{5}{2}}$$

$$= 27x(9 - x^2)^{-\frac{3}{2}}$$

$$= \frac{27x}{(9 - x^2)^{\frac{3}{2}}}$$

Hence $(9 - x^2)\dfrac{d^2y}{dx^2}$ gives us $\dfrac{27x}{(9 - x^2)^{\frac{3}{2}}}$

Let us now simplify $3x\dfrac{dy}{dx}$ as follows:

$$3x[9(9 - x^2)^{-\frac{3}{2}}] \qquad \text{(Note that } \frac{dy}{dx} = 9(9 - x^2)^{-\frac{3}{2}} \text{ as obtained above)}$$

$$= 27x(9 - x^2)^{-\frac{3}{2}}$$

$$= \frac{27x}{(9 - x^2)^{\frac{3}{2}}}$$

Hence $3x\dfrac{dy}{dx}$ also gives us $\dfrac{27x}{(9-x^2)^{\frac{3}{2}}}$

Therefore, $(9-x^2)\dfrac{d^2y}{dx^2} = 3x\dfrac{dy}{dx}$ as both sides give $\dfrac{27x}{(9-x^2)^{\frac{3}{2}}}$

14. Given that $y = \dfrac{x}{x-1}$, show that: $(x-1)\dfrac{d^2y}{dx^2} + 2\dfrac{dy}{dx} = 0$

Solution

$y = \dfrac{x}{x-1}$

Using quotient rule gives $\dfrac{dy}{dx}$ as follows:

$\dfrac{dy}{dx} = \dfrac{(x-1)(1) - x(1)}{(x-1)^2}$

$\quad = \dfrac{x-1-x}{(x-1)^2}$

$\dfrac{dy}{dx} = \dfrac{-1}{(x-1)^2}$

Using chain rule gives $\dfrac{d^2y}{dx^2}$ as follows:

$\dfrac{dy}{dx} = \dfrac{-1}{(x-1)^2}$

$\quad = -1(x-1)^{-2}$

$\dfrac{d^2y}{dx^2} = -2 \times -1(x-1)^{-2-1} \times \dfrac{d(x)}{dx}$

$\quad = 2(x-1)^{-3} \times 1$

$\dfrac{d^2y}{dx^2} = \dfrac{2}{(x-1)^3}$

From the question, let us now simplify $(x-1)\dfrac{d^2y}{dx^2} + 2\dfrac{dy}{dx}$ as follows:

$(x-1)\dfrac{d^2y}{dx^2} + 2\dfrac{dy}{dx}$

$= (x-1)\dfrac{2}{(x-1)^3} + (2)\dfrac{-1}{(x-1)^2}$

$= \dfrac{2}{(x-1)^2} - \dfrac{2}{(x-1)^2}$

$= 0$

This shows that $(x-1)\dfrac{d^2y}{dx^2} + 2\dfrac{dy}{dx}$ is equal to zero.

15. Find, with respect to x, the derivative of $\left(x - \dfrac{5}{x}\right)^3$

Solution

$$y = \left(x - \frac{5}{x}\right)^3$$

This can also be written as:

$$y = (x - 5x^{-1})^3$$

$$\frac{dy}{dx} = 3(x - 5x^{-1})^{3-1} \times \frac{d(x - 5x^{-1})}{dx} \qquad \text{(By use of chain rule)}$$

$$= 3(x - 5x^{-1})^2 \times 1 - (-1 \times 5x^{-1-1})$$

$$= 3(x - 5x^{-1})^2 \times 1 - (-5x^{-2})$$

$$= 3(x - 5x^{-1})^2 \times (1 + 5x^{-2})$$

$$= 3(x - 5x^{-1})^2(1 + 5x^{-2})$$

$$\frac{dy}{dx} = 3\left(x - \frac{5}{x}\right)^2 \left(1 + \frac{5}{x^2}\right)$$

16. Given that $\exp(2x^2 + 2y^2 - 16) = x + y$, find:

(a) $\dfrac{dy}{dx}$

(b) $\dfrac{dy}{dx}$ at $\left(\dfrac{1}{2}, \dfrac{1}{2}\right)$

Solution

$$\exp(2x^2 + 2y^2 - 16) = x + y$$

This can also be written as

$$e^{2x^2 + 2y^2 - 16} = x + y \qquad \text{(Note that } \exp x = e^x)$$

We now differentiate implicitly as follows:

$$\frac{d(2x^2 + 2y^2 - 16)}{dx} \times (e^{2x^2 + 2y^2 - 16}) = \frac{d(x)}{dx} + \frac{d(y)}{dx}$$

$$\left(4x + 4y\frac{dy}{dx}\right)(e^{2x^2 + 2y^2 - 16}) = 1 + \frac{dy}{dx}$$

Expanding the bracket gives:

$$4x(e^{2x^2 + 2y^2 - 16}) + 4y\frac{dy}{dx}(e^{2x^2 + 2y^2 - 16}) = 1 + \frac{dy}{dx}$$

Collecting terms in $\dfrac{dy}{dx}$ on the left hand side gives:

$$4y\frac{dy}{dx}(e^{2x^2 + 2y^2 - 16}) - \frac{dy}{dx} = 1 - 4x(e^{2x^2 + 2y^2 - 16})$$

Factorizing the left hand side gives:

$$\frac{dy}{dx}[4y(e^{2x^2 + 2y^2 - 16}) - 1] = 1 - 4x(e^{2x^2 + 2y^2 - 16})$$

Hence, $\dfrac{dy}{dx} = \dfrac{1 - 4x(e^{2x^2 + 2y^2 - 16})}{4y(e^{2x^2 + 2y^2 - 16}) - 1}$

Or, $\dfrac{dy}{dx} = \dfrac{-[4x(e^{2x^2 + 2y^2 - 16}) - 1]}{4y(e^{2x^2 + 2y^2 - 16}) - 1}$

(b) In order to find $\dfrac{dy}{dx}$ at $\left(\dfrac{1}{2} \ \dfrac{1}{2}\right)$, we simply substitute $x = \dfrac{1}{2}$ and $y = \dfrac{1}{2}$ into the expression for $\dfrac{dy}{dx}$ as follows:

$$\frac{dy}{dx} = \frac{-[4x(e^{2x^2 + 2y^2 - 16}) - 1]}{4y(e^{2x^2 + 2y^2 - 16}) - 1}$$

Since $x = \dfrac{1}{2}$ and $y = \dfrac{1}{2}$, this simplifies to give:

$$\frac{dy}{dx} \text{ at } \left(\frac{1}{2} \ \frac{1}{2}\right) = \frac{-[4\left(\frac{1}{2}\right)(e^{2x^2 + 2y^2 - 16}) - 1]}{4\left(\frac{1}{2}\right)(e^{2x^2 + 2y^2 - 16}) - 1}$$

$$= \frac{-[2(e^{2x^2 + 2y^2 - 16}) - 1]}{2(e^{2x^2 + 2y^2 - 16}) - 1}$$

The numerator cancels out the denominator to give:

$$\frac{dy}{dx} = \frac{-1}{1}$$

$$\frac{dy}{dx} = -1$$

17. If $y = x^3 - 2x^2 + 5$, show that: $x\dfrac{dy}{dx} - 3y - 2x^2 + 15 = 0$

Solution

$$y = x^3 - 2x^2 + 5$$

$$\frac{dy}{dx} = 3x^2 - 4x$$

Let us now simplify $x\dfrac{dy}{dx} - 3y - 2x^2 + 15$ as follows:

$$x\frac{dy}{dx} - 3y - 2x^2 + 15$$
$$= x(3x^2 - 4x) - 3y - 2x^2 + 15$$
$$= x(3x^2 - 4x) - 3(x^3 - 2x^2 + 5) - 2x^2 + 15 \quad \text{(Note that } x^3 - 2x^2 + 5 \text{ has been substituted for y)}$$
$$= 3x^3 - 4x^2 - 3x^3 + 6x^2 - 15 - 2x^2 + 15$$
$$= 0$$

Therefore, $x\dfrac{dy}{dx} - 3y - 2x^2 + 15 = 0$ as proven above.

18. Determine $\dfrac{d^2}{dx^2}\left(x\sin\dfrac{1}{x}\right)$

Solution

This means the second derivative of $x\sin\dfrac{1}{x}$

Let $y = x\sin\dfrac{1}{x}$

Or, $y = x\sin x^{-1}$

273

We now apply product rule to obtain $\dfrac{dy}{dx}$ as follows:

$$\frac{dy}{dx} = x\left[\frac{d(x^{-1})}{dx}\cos x^{-1}\right] + \sin x^{-1}\left[\frac{d(x)}{dx}\right]$$

$$= x(-x^{-2}\cos x^{-1}) + \sin x^{-1}(1)$$

$$= x(\frac{-1}{x^2}\cos x^{-1}) + \sin x^{-1}$$

$$= \frac{-x}{x^2}\cos x^{-1} + \sin x^{-1}$$

$$= \frac{-\cos x^{-1}}{x} + \sin x^{-1}$$

$$\frac{dy}{dx} = -\frac{1}{x}\cos\frac{1}{x} + \sin\frac{1}{x}$$

Or, $\dfrac{dy}{dx} = -x^{-1}\cos x^{-1} + \sin x^{-1}$

We now obtain $\dfrac{d^2y}{dx^2}$ as follows:

$$\frac{d^2y}{dx^2} = -x^{-1}\left[\frac{d(\cos x^{-1})}{dx}\right] + \cos x^{-1}\left[\frac{d(-x^{-1})}{dx}\right] + \frac{d(\sin x^{-1})}{dx}$$

$$= -x^{-1}\left[\frac{d(x^{-1})}{dx}(-\sin x^{-1})\right] + \cos x^{-1}(x^{-2}) + \left[\frac{d(x^{-1})}{dx}\cos x^{-1}\right]$$

$$= -x^{-1}[-x^{-2}(-\sin x^{-1})] + \cos x^{-1}(x^{-2}) - x^{-2}(\cos x^{-1})$$

$$= -x^{-1}(x^{-2}\sin x^{-1}) + x^{-2}\cos x^{-1} - x^{-2}\cos x^{-1}$$

$$= -x^{-1}(x^{-2}\sin x^{-1}) \quad \text{(Note that } x^{-2}\cos x^{-1} \text{ cancels out)}$$

$$= -x^{-3}\sin x^{-1}$$

$$\frac{d^2y}{dx^2} = -\frac{1}{x^3}\sin\frac{1}{x}$$

19. Given that $y = \dfrac{e^x + e^{-x}}{e^x - e^{-x}}$

(a) find $\dfrac{d^2y}{dx^2}$

(b) show that $\dfrac{d^2y}{dx^2} + 2y\dfrac{dy}{dx} = 0$

Solution

(a) $y = \dfrac{e^x + e^{-x}}{e^x - e^{-x}}$

By using quotient rule we obtain $\dfrac{dy}{dx}$ as follows:

$$\frac{dy}{dx} = \frac{e^x - e^{-x}\left[\frac{d(e^x + e^{-x})}{dx}\right] - \left(e^x + e^{-x}\left[\frac{d(e^x - e^{-x})}{dx}\right]\right)}{(e^x - e^{-x})^2}$$

$$= \frac{(e^x - e^{-x})(e^x - e^{-x}) - (e^x + e^{-x})(e^x + e^{-x})}{(e^x - e^{-x})^2}$$

$$= \frac{(e^x - e^{-x})^2 - (e^x + e^{-x})^2}{(e^x - e^{-x})^2}$$

Separating into fractions gives:

$$= \frac{(e^x - e^{-x})^2}{(e^x - e^{-x})^2} - \frac{(e^x + e^{-x})^2}{(e^x - e^{-x})^2}$$

$$\frac{dy}{dx} = 1 - \frac{(e^x + e^{-x})^2}{(e^x - e^{-x})^2}$$

We now apply quotient and chain rules to obtain $\frac{d^2y}{dx^2}$ as follows:

$$\frac{d^2y}{dx^2} = 0 - \frac{(e^x - e^{-x})^2[2(e^x + e^{-x})(e^x - e^{-x})] - (e^x + e^{-x})^2[2(e^x - e^{-x})(e^x + e^{-x})]}{[(e^x - e^{-x})^2]^2}$$

$$= -\left[\frac{2(e^x - e^{-x})^3(e^x + e^{-x}) - 2(e^x + e^{-x})^3(e^x - e^{-x})}{(e^x - e^{-x})^4}\right]$$

$$= \frac{-2(e^x - e^{-x})^3(e^x + e^{-x}) + 2(e^x + e^{-x})^3(e^x - e^{-x})}{(e^x - e^{-x})^4}$$

$$= \frac{2(e^x + e^{-x})^3(e^x - e^{-x}) - 2(e^x - e^{-x})^3(e^x + e^{-x})}{(e^x - e^{-x})^4} \qquad \text{(After rearranging the numerator)}$$

Separating into fractions gives:

$$= \frac{2(e^x + e^{-x})^3(e^x - e^{-x})}{(e^x - e^{-x})^4} - \frac{2(e^x - e^{-x})^3(e^x + e^{-x})}{(e^x - e^{-x})^4}$$

$$\frac{d^2y}{dx^2} = \frac{2(e^x + e^{-x})^3}{(e^x - e^{-x})^3} - \frac{2(e^x + e^{-x})}{e^x - e^{-x}}$$

(b) Let us simplify $\frac{d^2y}{dx^2} + 2y\frac{dy}{dx}$ as follows:

$$\frac{d^2y}{dx^2} + 2y\frac{dy}{dx}$$

$$= \frac{2(e^x + e^{-x})^3}{(e^x - e^{-x})^3} - \frac{2(e^x + e^{-x})}{e^x - e^{-x}} + 2\left(\frac{e^x + e^{-x}}{e^x - e^{-x}}\right)\left[1 - \frac{(e^x + e^{-x})^2}{(e^x - e^{-x})^2}\right]$$

Expanding bracket gives:

$$= \frac{2(e^x + e^{-x})^3}{(e^x - e^{-x})^3} - \frac{2(e^x + e^{-x})}{e^x - e^{-x}} + 2\left(\frac{e^x + e^{-x}}{e^x - e^{-x}}\right) - \frac{2(e^x + e^{-x})^3}{(e^x - e^{-x})^3}$$

$= 0$ (Since equal terms with opposite signs cancel out each other)

Therefore, $\frac{d^2y}{dx^2} + 2y\frac{dy}{dx}$ gives zero as proven above.

Or, $\frac{d^2y}{dx^2} + 2y\frac{dy}{dx} = 0$

20. Find the derivative of ln(tan2x)

<u>Solution</u>

y = ln(tan2x)

$$\frac{dy}{dx} = \frac{\frac{d(\tan 2x)}{dx}}{\tan 2x}$$

$$= \frac{2\sec^2 2x}{\tan 2x}$$

Or, $\dfrac{dy}{dx} = \dfrac{2\left(\frac{1}{\cos 2x}\right)\left(\frac{1}{\cos 2x}\right)}{\left(\frac{\sin 2x}{\cos 2x}\right)}$ (Note that sec2x = $\dfrac{1}{\cos 2x}$ and tan2x = $\dfrac{\sin 2x}{\cos 2x}$)

$$= \left(\frac{2}{\cos 2x}\right)\left(\frac{1}{\cos 2x}\right) \text{ x } \frac{\cos 2x}{\sin 2x}$$

$$\frac{dy}{dx} = \frac{2}{\sin 2x \cos 2x}$$ (Since cos2x cancels cos2x)

21. Given that y = $x^3 + 3x^2$, determine $2\dfrac{dy}{dx} - x\dfrac{d^2y}{dx^2}$

Solution

$$y = x^3 + 3x^2$$

$$\frac{dy}{dx} = 3x^2 + 6x$$

$$\frac{d^2y}{dx^2} = 6x + 6$$

Hence, $2\dfrac{dy}{dx} - x\dfrac{d^2y}{dx^2}$ is simplified as follows:

$$2(3x^2 + 6x) - x(6x + 6)$$

$$= 6x^2 + 12x - 6x^2 - 6x$$

$$= 6x$$

Therefore, $2\dfrac{dy}{dx} - x\dfrac{d^2y}{dx^2} = 6x$

22. Find $\dfrac{dy}{dx}$ if y = $3(3x + \sqrt{x})^2$

Solution

$$y = 3(3x + \sqrt{x})^2$$

Applying chain rule gives:

$$\frac{dy}{dx} = 3 \text{ x } 2\,(3x + x^{\frac{1}{2}})^{2-1} \text{ x } \frac{d(3x + x^{\frac{1}{2}})}{dx}$$ (Note that $\sqrt{x} = x^{\frac{1}{2}}$)

$$= 6(3x + x^{\frac{1}{2}}) \text{ x } (3 + \frac{1}{2}x^{-\frac{1}{2}})$$

$$= 6(3x + x^{\frac{1}{2}})\left(3 + \frac{1}{2x^{\frac{1}{2}}}\right)$$

$$= 6(3x + \sqrt{x})\left(3 + \frac{1}{2\sqrt{x}}\right)$$

Expanding the bracket gives:

$$= 6\left(9x + \frac{3x}{2\sqrt{x}} + 3\sqrt{x} + \frac{1}{2}\right)$$

$$= 6\left(9x + \frac{3}{2}\sqrt{x} + 3\sqrt{x} + \frac{1}{2}\right) \qquad \text{(Note that } \frac{3x}{2\sqrt{x}} = \frac{3}{2}x^{1-\frac{1}{2}} = \frac{3}{2}x^{\frac{1}{2}} = 3\sqrt{x}\text{)}$$

$$= 6\left(9x + \frac{9}{2}\sqrt{x} + \frac{1}{2}\right)$$

$$\frac{dy}{dx} = 54x + 27\sqrt{x} + 3$$

23. If $y = \dfrac{5x^2 + 7}{x^4}$,

(a) find $\dfrac{d^2y}{dx^2}$

(b) show that $x^2\dfrac{d^2y}{dx^2} + 7x\dfrac{dy}{dx} + 8y = 0$

Solution

(a) $y = \dfrac{5x^2 + 7}{x^4}$

$$= \frac{5x^2}{x^4} + \frac{7}{x^4}$$

$$= \frac{5}{x^2} + \frac{7}{x^4}$$

$$y = 5x^{-2} + 7x^{-4}$$

$$\frac{dy}{dx} = -10x^{-3} - 28x^{-5}$$

Similarly,

$$\frac{d^2y}{dx^2} = 30x^{-4} + 140x^{-6}$$

(b) Let us now simplify $x^2\dfrac{d^2y}{dx^2} + 7x\dfrac{dy}{dx} + 8y$ by substituting appropriately as follows:

$$x^2(30x^{-4} + 140x^{-6}) + 7x(-10x^{-3} - 28x^{-5}) + 8(5x^{-2} + 7x^{-4})$$

$$= 30x^{-2} + 140x^{-4} - 70x^{-2} - 196x^{-4} + 40x^{-2} + 56x^{-4}$$

$$= 30x^{-2} + 40x^{-2} - 70x^{-2} + 140x^{-4} + 56x^{-4} - 196x^{-4}$$

$$= 0$$

Therefore $x^2\dfrac{d^2y}{dx^2} + 7x\dfrac{dy}{dx} + 8y$ gives zero as obtained above.

Or, $x^2\dfrac{d^2y}{dx^2} + 7x\dfrac{dy}{dx} + 8y = 0$

24. If $y = (2x + 5)^4 + \dfrac{x - 1}{2x - 1}$ find $\dfrac{dy}{dx}$.

Solution

$$(2x + 5)^4 + \frac{x - 1}{2x - 1}$$

We now use chain rule and quotient rule as follows:

$$\frac{dy}{dx} = 4(2x+5)^{4-1} \times \frac{d(2x+5)}{dx} + \frac{(2x-1)\frac{d(x-1)}{dx} - \left[(x-1)\frac{d(2x-1)}{dx}\right]}{(2x-1)^2}$$

$$= 4(2x+5)^3(2) + \frac{(2x-1)(1) - \left[(x-1)(2)\right]}{(2x-1)^2}$$

$$= 8(2x+5)^3 + \frac{(2x-1) - 2(x-1)}{(2x-1)^2}$$

$$= 8(2x+5)^3 + \frac{2x-1-2x+2}{(2x-1)^2}$$

$$\frac{dy}{dx} = 8(2x+5)^3 + \frac{1}{(2x-1)^2}$$

Exercise 24

1. If $y = x^2(3x^4 - 5)^3$ find $\frac{dy}{dx}$.

2. Given that $f(x) = 2x^5 - x^4 + 2x^3 + x^2 - 3x + 4$ find $f'(-2)$

3. A function $f(x)$ is given by $f(x) = \frac{\sqrt{(3x^2 - 1)^3}}{x^2}$. Find:

(a) the derivative of $f(x)$

(b) the gradient of $f(x)$ at the point $(1, -2)$

4. If $3xy^3 - y^2 - 4x^3 = 5y$, find:

(a) the derivative of the expression

(b) the gradient at $(-1, 1)$

5. Find the derivative of $(2x^2 + y^2)^2 = 0$

6. If $y = (x^2 + 5)^3(x^3 - 1)$, find $\frac{dy}{dx}$.

7. Find the derivative of $e^{\sin x + \tan x}$

8. Find the derivative of $\frac{x^2}{\sin^2 x}$

9. Differentiate with respect to x: $\ln(\sin^2 x + \cos 3x)$

10. If $y = a^{\cos 5x}$ find $\frac{dy}{dx}$.

11. Given that $y = x^2 e^{-3x}$ evaluate $\frac{d^2y}{dx^2} + \frac{dy}{dx} - y$

12. Differentiate with respect to x: $\sin\left(\frac{1 - 2x^3}{x^2}\right)$

13. If $y = \frac{2x-1}{x^2}$, find $x^4\frac{d^2y}{dx^2} - 3(2x-1)\frac{dy}{dx}$

14. Given that y $=5e^{-2x} + 3e^x$, evaluate $\dfrac{d^2y}{dx^2} + \dfrac{dy}{dx} - 2y$

15. Find the derivative of $\left(\dfrac{1}{x^2} - \dfrac{2}{x}\right)^5$

16. Given that $e^{x^3 - y^3} = xy$, find:

(a) $\dfrac{dy}{dx}$

(b) $\dfrac{dy}{dx}$ at (1, −1)

17. If y = sin(sinx), evaluate $\dfrac{d^2y}{dx^2} + \tan x\,\dfrac{dy}{dx} + y\cos^2 x$

18. Determine $\dfrac{d^2}{dx^2}\left(x^2\cos\dfrac{1}{x^2}\right)$

19. Given that $y = \dfrac{1 + e^{-x}}{1 - e^{-x}}$

(a) find $\dfrac{d^2y}{dx^2}$

(b) evaluate $\dfrac{d^2y}{dx^2} - \dfrac{dy}{dx} + y$

20. Find the derivative of $\ln(\sec^2 x)$

21. Given that y = x + tanx, determine $\cos^2 x\,\dfrac{d^2y}{dx^2} - 2y + 2x$

22. Find $\dfrac{dy}{dx}$ if y = $\left(3x + \dfrac{\sqrt{5x}}{2}\right)^3$

23. If $y = \dfrac{x^3 + 2x^2 - 5x - 3}{x^2}$, find $\dfrac{d^2y}{dx^2}$

24. If y = $(x^3 + 1)^3 + \dfrac{5x - 2}{x^2 + 1}$ find $\dfrac{dy}{dx}$.

25. Given that y = 2cos5x + 7sin5x, determine $\dfrac{d^2y}{dx^2} + 25y$

ANSWERS TO EXERCISES

Exercise 1

1. (a) 2   (b) $-\dfrac{5}{4}$ (c) $\dfrac{15}{4}$ (d) 0.53

2. B = 6   3. (a) $\sqrt{2}$ (c) $\sqrt{13}$ (d) $\sqrt{26}$ (d) 5

4. (a) (0, 5)   (b) (2, 1)   (c) (1, 4)   (d) (1, 2)

5. (a) $\left(\dfrac{5}{4}, \dfrac{5}{4}\right)$ (b) (4, 5)   (c) (1, –5)   (d) $(7, \dfrac{7}{2})$ (e) $\left(\dfrac{5}{4}, \dfrac{5}{4}\right)$

6. 2 : 3   7. 3 : 1   8. 1 : 2   9. –2 ; 13 or 2 : 13 externally

10. P = (0, 20 and Q = (2, 3)   11. $\dfrac{5}{\sqrt{5}}$ or $\sqrt{5}$ 12. $\dfrac{9}{\sqrt{13}}$ or $\dfrac{9\sqrt{13}}{13}$ 13. –3 and 108.4°

14. –2.5 and 111.8°   15. (a) Parallel   (a) Perpendicular   (c) Perpendicular

(d) Perpendicular   16. (a) $-\dfrac{5}{2}$ (b) 3   (c) 5   17. $\dfrac{3\sqrt{10}}{2}$ 18. $\dfrac{9}{\sqrt{29}}$ or $\dfrac{9\sqrt{29}}{29}$

19. 4.4°   20. 125.54°   21. 3.37° and 176.63°   22. (a) $y = x + 5$ (b) $2y = x + 2$

23. (a) $y + 3x = 21$ (b) $x + y = 10$ 24. (a) x - intercept $= \dfrac{3}{2}$, y - intercept $= -\dfrac{3}{7}$

(b) x - intercept $= \dfrac{8}{5}$, y - intercept = 8   25. $5y - 3x + 18 = 0$ 26. $x - y = 2$

27. $2y + x + 3 = 0$ 28. $3x - y - 26 = 0$ 29. (1, –1)   30. (2, 3)

31. $y + 2x = 20$ 32. (a) $2y + x = 3$ (b) $y - 2x + 11 = 0$ 33. $x - 1 = 0$

34. $\dfrac{9}{2\sqrt{5}}$ or $\dfrac{9\sqrt{5}}{10}$ 35. $\dfrac{13}{\sqrt{5}}$ or $\dfrac{13\sqrt{5}}{5}$

Exercise 2

1. (a) $y + 3x - 11 = 0$ (b) P = (11, 0) and Q = (0, 11)   (c) 8 square units

2. (a) m = (0, –3) and N = (1, 2)   (b) 7 square units   3. 36 square units

4. 16 square units   5. (a) $\left(-\dfrac{9}{5}, \dfrac{13}{5}\right)$ (b) $y + 2x = -1$ (c) 2 square units

6. 4 square units   7. $\dfrac{1}{4}$ Square unit

Exercise 3

1. $4x + 14y + 13 = 0$ 2. $2x^2 + 2y^2 - 8y + 39 = 0$ 3. $4x^2 - 16x + 28y - 5 = 0$

4. $y - 6x - 17 = 0$ 5. $6x^2 + 6y^2 - 12x - 48y - 173 = 0$ 6. $2x - 9y - 8 = 0$

7. $x^2 + y^2 - 3x + y + 4 = 0$ 8. $7y - x - 2 = 0$

Exercise 4

1. $x^2 + y^2 = 81$ 2. $x^2 + y^2 + 2x + 2y - 23 = 0$ 3. (a) $\sqrt{20}$

(b) $x^2 + y^2 - 4x - 10y + 9 = 0$ 4. (a) $\dfrac{9\sqrt{2}}{2}$ (b) $x^2 + y^2 + 3x - 3y - 36 = 0$

5. (a) (-2, 3) (b) $\sqrt{5}$ units 6. Center = $\left(\frac{3}{2}, -\frac{7}{2}\right)$, r = 4.45 units

7. $x^2 + y^2 - 10x + 6y + 9 = 0$ 8. (a) 10 units (b) $x^2 + y^2 - 22x - 4y + 25 = 0$

9. $4y - 2x - 21 = 0$ 10. (a) Yes (b) No (x^2 and y^2 have different coeeficient)

(c) No (Same reason in (b) above) (d) Yes (e) Yes (f) Yes

(g) No (Presence of y^3 and xy (h) No ($-y^2$ means a coefficient of -1 for y^2) (i) Yes

11. It lies on the circle 12. (0, 3) and (3, 0) 13. $7x^2 + 7y^2 - 46x + 82y + 110 = 0$

14. (a) $8x^2 + 8y^2 - 36x - 18y + 45 = 0$ (b) Center = $\left(\frac{9}{4}, \frac{9}{8}\right)$, r = $\frac{\sqrt{45}}{8}$ or $\frac{3\sqrt{5}}{8}$

15. $x^2 + y^2 - 4x + 10y + 4 = 0$ 16. K = -44 17. (a) $\frac{5}{3}\pi$ square units

(b) Diameter = 2.58 units 18. $y + 2x + 1 = 0$ 19. $y - 2x - 3 = 0$ 20. $9y - x + 26 = 0$

21. $12x - y + 25 = 0$ 22. $y - x + 1 = 0$ 23. $\sqrt{134}$ units 24. $\sqrt{21}$ units

25. $x^2 + y^2 + 4x + 2y + 4 = 0$ 26. $x^2 + y^2 - 6x - 10y + 9 = 0$ 27. $x^2 + y^2 - 20x - 14y + 49 = 0$

28. $x^2 + y^2 - 10x - 10y + 25 = 0$ 29. (a) r = 2.5 units (b) $4x^2 + 4y^2 - 24x - 40y + 111 = 0$

30. (2, 1) 31. $y - x + 2.47 = 0$ and $y - x - 6.47 = 0$ 32. (0, 2) 33. (1, 2) 34. 5

35. P($-3, -2$) and Q($-17, -16$) 36. 4 units 37. $3x^2 + 3y^2 - 13x - 11y + 20 = 0$

38. (1, -10)

Exercise 5

1. Vertex = (0, 0), focus = (6, 0), directrix is: $x = -6$ 2. (a) $y^2 = -20x$ (b) $x = 5$

(c) 20 units 3. (a) (0, 8) (b) $y = -8$ (c) Upwards 4. (a) (0, -10) (b) $y = 1$

5. (a) ($-3, 5$) (b) ($-1, 5$) (c) $x = -5$ 6. (a) ($-1, 3$) (b) $-1, 60$ (c) $y = 0$

7. $2y + 4x - 5 = 0$ 8. (a) $y - 2x - 4 = 0$ (b) $2y + x - 18 = 0$ 9. $x + y - 5 = 0$

10. $y^2 + 10y - 6x + 28 = 0$ 11. $x^2 - 4x + 4y - 28 = 0$

12. $y^2 + 4y - 12x + 16 = 0$, directrix is $x = -2$ 13. (a) $x^2 - 10x + 20y - 35 = 0$ (b) $y = 8$

14. $x^2 - 4x - 16y + 20 = 0$ (b) (2, 5) 15. (a) $x^2 - 10x + 20y - 35 = 0$ (b) ($-5, -4$)

Exercise 6

1. $V_1 = (0, 6)$, $V_2 = (0, -6)$, $V_3 = (4, 0)$, $V_4 = (-4, 0)$, $F_1 = (0, \sqrt{20})$, $F_2 = (0, -\sqrt{20})$

2. $V_1 = (10, 0)$, $V_2 = (-10, 0)$, $V_3 = (0, 8)$, $V_4 = (0, -8)$, $F_1 = (6, 0)$, $F_2 = (-6, 0)$

3. a = 5, b = 2, c = $\sqrt{21}$ 4. Major axis = $6\sqrt{5}$ units, Minor axis = $4\sqrt{5}$ units

5. $\frac{x^2}{36} + \frac{y^2}{100} = 1$ 6. $\frac{x^2}{289} + \frac{y^2}{64} = 1$ 7. (a) e = $\frac{4}{5}$ (b) $\frac{18}{25}$ units (c) 15π square units

8. $\frac{x^2}{169} + \frac{y^2}{25} = 1$ 9. $\frac{x^2}{100} + \frac{y^2}{25} = 1$ 10. $\frac{(x+2)^2}{10} + \frac{4\left(y - \frac{5}{2}\right)^2}{49} = 1$ or $\frac{(x+2)^2}{10} + \frac{(2y-5)^2}{49} = 1$

11. $\frac{(x-2)^2}{100} + \frac{(y-3)^2}{51} = 1$ 12. (a) (2, 3) (b) $V_1 = (12, 3)$, $V_2 = (-8, 3)$, $V_3 = (2, \sqrt{51} + 3)$,

$V_4 = (2, -\sqrt{51} + 3)$ (c) $F_1 = (9, 3)$, $F_2 = (-5, 3)$ 13. (a) (1, 2) (b) $V_1 = (5, 2)$, $V_2 = (-3, 2)$,

(c) $F_1 = (2\sqrt{3} + 1, 2)$, $F_2 = (-2\sqrt{3} + 1, 2)$ 14. $x = -5$ (15) $x = 0$

Exercise 7

1. $V_1 = (2, 0)$ and $V_2 = (-2, 0)$, $F_1 = (\sqrt{29}, 0)$ and $F_2 = (-\sqrt{29}, 0)$ 2.(a) 12 units

(b) 20 units   (c) 1.33 units   (d) $(0, \sqrt{136})$ and $(0, -\sqrt{136})$ (e) $(0, 6)$ and $(0, -6)$

(f) $\dfrac{50}{9}$ units   (g) $y = -\dfrac{3}{5}$ and $y = \dfrac{3}{5}$ 3. $e = \dfrac{\sqrt{5}}{2}$ 4. $\dfrac{y^2}{9} - \dfrac{x^2}{4} = 1$ 5. $\dfrac{x^2}{64} - \dfrac{y^2}{225} = 1$

6. (a) $e = 1.86$ (b) 78.125 units   7. $\dfrac{16x^2}{225} - \dfrac{16y^2}{175} = 1$ 8. $\dfrac{2x^2}{5} - \dfrac{3y^2}{5} = 1$

9. $\dfrac{(y+4)^2}{4} - \dfrac{(x-1)^2}{12} = 1$ 10. $\dfrac{(x+2)^2}{9} - \dfrac{(y-2)^2}{27} = 1$ 11. (a) $(2, -2)$

(b) $V_1 = (4, -2)$, $V_2 = (0, -2)$, $V_3 = (2, 1)$, $V_4 = (2, -5)$ (c) $F_1 = (\sqrt{13} + 2, -2)$, $F_2 = (-\sqrt{13} + 2, -2)$

12. (a) $(-2, 1)$ (b) $V_1 = (-2, 3)$, $V_2 = (-2, -1)$ (c) $F_1 = (-2, \sqrt{29} + 1)$, $F_2 = (-2, -\sqrt{29} + 1)$

13. $4x\sqrt{2} - 6y - 12 = 0$ 14. $8x - 5y = 100$ 15. $\dfrac{x^2}{64} - \dfrac{y^2}{225} = 1$

16. (a) Hyperbola   (b) Hyperbola   (c) Parabola   (d) Circle   (e) Ellipse   (f) Circle

(g) Circle   (h) Parabola   (i) Ellipse   (j) Hyperbola

Exercise 8

1. 9   2. $-\dfrac{8}{5}$ 3. 9   4. 14   5. -14 6. $\dfrac{2}{5}$ 7. $\dfrac{1}{3}$ 8. -1 9. -15

10. 10   11. $\dfrac{1}{4}$ 12. 27   13. 12   14. $\dfrac{1}{10}$ 15. $\dfrac{3}{7}$ 16. Continuous

17. Continuous   18. Discontinuous   19. Not continuous (Discontinuous

20. Not continuous   21. Continuous   22. Not continuous   23. Not continuous

24(a) 8   (b) 5   26. $\dfrac{7}{2}$ 27. $\dfrac{1}{3}$ 28. 1   29. $-\dfrac{1}{2}$ 30. $2\dfrac{1}{4}$

Exercise 9

1. 2   2. $2x$ 3. $-3x^{-4}$ or $-\dfrac{3}{x^4}$ 4(a) $10x + 5h$ (b) $10x$

5(a) $27x^2 + 27xh + 9h^2$ (b) $27x^2$ 6(a) $\dfrac{2x^3 + 3x^2\Delta x + x(\Delta x)^2 + 2}{x(x + \Delta x)}$ (b) $2x + \dfrac{2}{x^2}$

7. $6x - 10$ 8. $1 - \dfrac{3}{x^2}$ 9. $5 - 6x$ 10. $2 + \dfrac{1}{5} = 2\dfrac{1}{5}$

Exercise 10

1(a) $40x^4$ (b) $2x^4$ (c) $\dfrac{1}{3x^{\frac{2}{3}}}$ or $\dfrac{1}{3\sqrt[3]{x^2}}$ (d) $\dfrac{1}{x^{\frac{6}{7}}}$ or $\dfrac{1}{\sqrt[7]{x^6}}$ (e) $-\dfrac{5}{8x^{\frac{13}{8}}}$ or $-\dfrac{5}{8\sqrt[8]{x^{13}}}$

(f) $-\dfrac{5}{x^{\frac{7}{2}}}$ or $-\dfrac{5}{\sqrt{x^7}}$ 2. (a) $10x^4 - 12x^3 - 12x^2 + 10x - 6$ (b) $7x^6 + 8x^3 + \dfrac{3}{x^2}$

(c) $18x^5 - 4x^3 - 5 - \dfrac{2}{x^2} + \dfrac{3}{x^4}$ (d) $\dfrac{5}{4\sqrt[4]{x^3}} - \dfrac{5}{3\sqrt[3]{2x^4}}$ 3. $8(2x - 5)^3$ 4. $\dfrac{-6}{(x^3 - 7)^3}$

5. $\dfrac{6x^2+7}{2(2x^3+7x)^{\frac{1}{2}}}$ 6. $5(21x^2-2x)(7x^3-x^2+3)^4$ 7. $\left(9+\dfrac{2}{x^2}\right)\left(3x-\dfrac{2}{3x}\right)^2$

8. $1+\dfrac{9(6x-1)}{(3x^2-x-10)^2}$ 9. $\dfrac{-4x^3}{\sqrt{1-2x^4}}$ 10. $\dfrac{-5x^2}{\sqrt[3]{(5x^3-1)^4}}$

Exercise 11

1. $12x-1$ 2. $12x^3+50x$ 3. $\dfrac{15x+30}{2\sqrt{3+x}}$ 4. $(7x^2+30-7)(x^2-7)^2$

5. $48x^3+24x^2-6x-20$ 6. $\dfrac{\sqrt{2}\,[(3x^2-1)^3+36x^2(9x^4-6x^2+1)]}{2\sqrt{x}}$

7. $\dfrac{15-7x}{4(x+3)^{\frac{1}{4}}}$ or $\dfrac{15-7x}{4\sqrt[4]{x+3}}$ 8. $12x^3+9x^2-46x-5$ 9. $15x^4-12x^3-3x^2+6x-2$

10. $\dfrac{12x+1}{2}$ 11. $\dfrac{9x^{\frac{7}{2}}}{2}$ or $\dfrac{9\sqrt{x^7}}{2}$ 12. $\dfrac{4x^3(7x-33)}{3\sqrt[3]{2x-11}}$

13. $-30x^5-175x^4+20x^3+102x^2-14x+1$ 14. $75x^4+28x^3+3x^2+28x-7$

15. $-\dfrac{2}{x^2}+\dfrac{12}{x^5}-\dfrac{25}{x^6}$

Exercise 12

1. $\dfrac{x^2-2x+4}{(x-1)^2}$ 2. $\dfrac{8x^2+24x-3}{(2x+3)^2}$ 3. $\dfrac{42x}{(3x^2+1)^2}$ 4. $\dfrac{1}{(1-x)^{\frac{3}{2}}\sqrt{x+1}}$

5. $\dfrac{4(x^3-2)(x^3+1)}{x^3}$ 6. $\dfrac{-6(x^2+1)}{x^4\sqrt{3x^2+2}}$ 7. $\dfrac{-8x^2+12x+24}{3(2x+1)^3(x^2-x-4)^{\frac{2}{3}}}$ 8. $\dfrac{10x^3-2x^2-x+4}{2(2-x)^{\frac{3}{2}}}$

9. $\dfrac{-6x}{(1-3x^2)^2}$ 10. $\dfrac{4}{(x-2)^2}$ or $\dfrac{4}{(2-x)^2}$

Exercise 13

1. $\dfrac{3t}{2}$ 2. $\dfrac{2(4-t^3)^2}{3t}$ 3. $\dfrac{2r}{3(l+2r)}$ 4. $\dfrac{2t-1}{10t-1}$ 5. $\dfrac{-2m(v-u)}{(u+v)t^2}$ 6. $\dfrac{5(t^4+1)^2}{3(t^2-1)^2}$

7. $\dfrac{1-12t^2}{2t}$ 8. $\dfrac{15t^2}{2}$ 9. $\dfrac{8}{r}$ 10. $\dfrac{2s^3}{3}$

Exercise 14

1. $\dfrac{-15x^2}{3-2y}$ or $\dfrac{15x^2}{2y-3}$ 2. $\dfrac{-4x}{9y^2}$ 3. $\dfrac{3y^2+12x^2-y}{x-6xy}$ 4. $\dfrac{3x^2-y^2-4x}{2xy}$ 5. $\dfrac{x^2}{y}$

6. $\dfrac{4xy}{2x^2+5}$ 7. $\dfrac{-2xy}{x^2+6y^2-1}$ 8. $\dfrac{4y}{10y-4x-1}$ 9. $\dfrac{2xy^2-1}{2y(x^2-1)}$ 10. $\dfrac{-5x^4y^2}{3}$

Exercise 15

1. $2\sec^2 2x$ 2. $-\dfrac{1}{5}\sin\dfrac{1}{5}x$ 3. $-50\sin 5x$ 4. $3\sin^2 x\cos x$ 5. $2\sec^2 x\tan x$

6. $60x^4\sin^3 3x^5\cos 3x^5$ 7. $-12x\cot 6x^2\csc 6x^2$ 8. $10x^4\sec 2x^5\tan 2x^5$ 9. $6x\sec^2 3x^2$

10. $-x(3x\sin3x - 2\cos3x)$ 11. $\dfrac{6x\cos 2x - 9\sin 2x}{x^4}$ 12. $48x^3\sec x^4\tan x^4$

13. $\dfrac{-5\tan x\cos 5x - \sec^2 x(2 - \sin 5x)}{\tan^2 x}$ 14. $\dfrac{\sin^2 x(3x\cos x - \sin x)}{2x^2}$

15. $\dfrac{-(2x+6)\operatorname{cosec}^2 2x - \cot 2x)}{(x+3)^2}$ 16. $\dfrac{-\sin\sqrt[3]{3}}{5x^2}$ 17. $\dfrac{2[x - (x^2- 3)\tan 2x]}{\sec 2x}$

18. $6x\cos3x^2 - 18x^3\sin3x^2$ 19. $-\cos x(2\cot x\sin x + \cos x\operatorname{cosec}^2 x)$

20. $\dfrac{2(\sin^2 x+ \sin^2 x)}{(\sin x+\cos x)^2}$ or $\dfrac{2}{(\sin x+\cos x)^2}$ 21. $\dfrac{4\sec 4x\sin (4x-2) + \tan 4x\cos (4x-2)}{\cos^2 (4x-2)}$

22. $-3\sin3x - x^2\sec x\tan x - 2x\sec x$ 23. $27x^2\cos x^3\sin^8 x^3$ 24. $\dfrac{3\cos 6\sqrt{x}}{\sqrt{x}}$

25. $3\cos x^3 - \dfrac{2\sin x^3}{x^3}$ 26. $-30x^4\cos^2 2x^5\sin 2x^2$ 27. $-10\sin2x\sin10x - 2\cos2x\cos10x$

28. $6x^2\cos x^4 - \dfrac{3\sin x^4}{2x^2}$ 29. $\dfrac{(10x - 5)\sec^2 5x - 2\tan 5x}{(2x-1)^2}$ 30. $\dfrac{\sec^2 x\tan x}{x} - \dfrac{\tan^2 x}{2x^2}$

Exercise 16

1. $\dfrac{1}{3\sqrt[3]{7y^2}}$ 2. $\dfrac{3y^2}{2\sqrt{y^3-1}}$ 3. $\dfrac{1}{5\sqrt[5]{2(y+3)^4}}$ 4. $\dfrac{1}{\sqrt[3]{2(3y+27)^2}}$ 5. $\dfrac{-1}{(y-2)^2}$

6. $\dfrac{-3}{2y^2\sqrt{\frac{3}{y}+5}}$ 7. $2y^3$ 8. $5y^4$ 9. $\dfrac{-5}{(y-4)^2}$ 10. $\dfrac{-1}{3y^2\sqrt[3]{(\frac{1}{y}+8)^2}}$

Exercise 17

1. $\dfrac{2}{\sqrt{1-4x^2}}$ 2. $\dfrac{-1}{\sqrt{1-x^2}}$ 3. $\dfrac{-6x}{9x^4+1}$ 4. $\dfrac{-4x^3}{x^8+1}$ 5. $3\sec^{-1}x + \dfrac{3}{\sqrt{x^2-1}}$

6. $2x + \dfrac{3}{x^2+1}$ 7. $\dfrac{1}{(x+5)^2+1}$ or $\dfrac{1}{x^2+10x+26}$ 8. $\dfrac{1}{x^2\sqrt{1-\frac{1}{x^2}}}$ 9. $\dfrac{-1}{5\sqrt{1-y^2}}$

10. $\dfrac{1}{2y\sqrt{y-1}}$ 11. $5\tan^{-1}3x + \dfrac{15x}{9x^2+1}$ 12. $2x + \dfrac{20x^4}{\sqrt{1-x^{10}}}$ 13. $\dfrac{-3}{x\sqrt{x^6-1}}$

14. $\dfrac{1}{2(y^2+1)}$ 15. $\dfrac{1}{15y^{\frac{2}{3}}\sqrt{1-y^{\frac{2}{3}}}}$

Exercise 18

1. $3\cosh3x - 2\sinh x$ 2. $-5x\operatorname{sech}x(x\tanh x - 2)$ 3. $\dfrac{2(x\cosh 2x - \sinh 2x)}{3x^3}$

4. $6\cosh3x\sinh3x$ 5. $60x^4\cosh^2 4x^5\sinh4x^5$ 6. $-10x^4\operatorname{cosech}2x^5$

7. $-12x\operatorname{sech}^3 2x^2\tanh2x^2$ 8. $x(5x\sinh5x + 2\cosh5x)$

9. $\dfrac{3\operatorname{cosech}^2 3x \tanh 5x + 5\coth 3x\operatorname{sech}^2 5x}{\coth^2 3x}$ 10. $\dfrac{2\cosh^2 x(3x\sinh x - \cosh x)}{3x^2}$

285

Exercise 19

1. $\frac{1}{x}\log_a e$ 2. $\frac{3x^2}{x^3+5}\log_a e$ 3. $\frac{36x^2}{2x^3-5}\log_a e$ 4. $\frac{2}{3x}\log_a e$ 5. $\frac{4x}{(x^2+1)(1-x^2)}\log_a e$

or $\frac{-4x}{(x^2+1)(x^2-1)}\log_a e$ or $\frac{-4x}{(x^4-1)}\log_a e$ 6. $\frac{15x^2}{5x^3-1}\log_5 e$ 7. $\frac{-2}{x}\log_2 e$ 8. $\frac{2x}{x^2+3}$

9. $\frac{3\ln^2 x}{x}$ 10. $\frac{5x^4}{2x^5-1}$ 11. $2x^3(4\ln x+1)$ 12. $\frac{-28}{3-7x}$ or $\frac{28}{7x-3}$

13. $6x\ln(4x^3+1)+\frac{36x^4}{4x^3+1}$ 14. $\frac{2-4\ln x}{x^3}$ 15. $\frac{-30x}{1-5x^2}$ or $\frac{30x}{5x^2-1}$

Exercise 20

1. $10a^{2x}\log_e a$ 2. $a^{5x^2-x}\ln a(10x-1)$ 3. $a^{3x}x^3(3x\ln a+4)$

4. $-2e^{-2x}-3e^{-x}$ or $-e^{-2x}(2+3e^x)$ 5. $\frac{\sqrt{3}\,e^{\sqrt{3x}}}{\sqrt{x}}$ 6. $6x^2e^{x^2}(2x^2+3)$

7. $\frac{5\ln a\,(a^{10x}-1)}{a^{5x}}$ 8. $\frac{6e^{4x}(2x\ln 2x^2+1)}{x}$ 9. $\frac{3x\sqrt{e^{5x}}(5x+4)}{2a}$ 10. $x^3(\log_{10}e+4\log_{10}7x)$

11. $3\ln 6(6^{3x})$ 12. $2x-3(e^{x^2-3x})$ 13. $\frac{-(5x+3)e^{\frac{3}{x}}}{5x^7}$ 14. $6e^{-3x}$

15. $3(2x+1)(x^2+x)^2e^{(x^2+x)^3}$ 16. $2^{x^2}2x\ln 2$ 17. $2a^{2x}\ln a$ 18. $\frac{e^x(x\ln 10x^3+3)}{x}$

19. $\frac{\sqrt{5}(4x+1)e^{2x}}{2\sqrt{x}}$ 20. $2x^4(4\log_{10}e+5\log_{10}5x^4)$

Exercise 21

1. $x^{2x}(2+2\ln x)$ 2. $\frac{3x^{\ln 2x}(\ln 2x+\ln x)}{x}$ 3. $\ln(1-4x^2)-\frac{8x^2}{1-4x^2}$

4. $(6x^x-5)^{3x}\left[3\ln(6x^2-5)+\frac{36x^2}{6x^2-5}\right]$ 5. $\frac{(2x^3+1)(x-2)^3}{3x^2(x^3-1)^2}\left(\frac{6x^2}{2x^3+1}+\frac{3}{x-2}-\frac{2}{x}-\frac{6x^2}{x^3-1}\right)$

6. $\frac{(2x-1)(x^2-2)}{(1-x)(x-3)^2}\left(\frac{2}{2x-1}+\frac{2x}{x^2-2}+\frac{1}{1-x}-\frac{2}{x-3}\right)$

7. $15x^3x^{3x^4}(4\ln x+1)$ or $15x^{3x^4+3}(4\ln x+1)$ 8. $3e^{3x}(e^{e^{3x}})$ or $3e^{e^{3x}+3x}$

9. $2^{e^x}e^x\ln 2$ 10. $\frac{x^{\ln(x^2-4x)}[(x-4)\ln(x^2-4x)+(2x-4)\ln x]}{x(x-4)}$ 11. $\ln(2x^2+10)+\frac{4x^2}{2x^2+10}$

12. $(3x^2-8)^{2x}\left[2\ln(3x^3-8)+\frac{18x^3}{3x^3-8}\right]$ 13. $\frac{(x+3)^2(x-3)^2}{x^3-1}\left(\frac{2}{x+3}+\frac{2}{x-3}-\frac{3x^2}{x^3-1}\right)$

14. $15x^{(\ln x)^2}(\ln x)^2$ 15. $30x^{3x^2}x(2\ln x+1)$ or $30x^{3x^2+1}(2\ln x+1)$

Exercise 22

1. $\frac{3}{5x^2}$ 2. $\frac{5xe^{5x}}{2}$ 3. $\frac{-2x\sin x^2}{\cos x}$ 4. $\frac{x(6x-5)}{x-1}$ 5. $\frac{14x^2}{3}$ 6. $\frac{2a^x\ln a}{e^x}$

7. $\frac{-2\tan x\sec^2 x}{5\sin 5x}$ 8. $\frac{1}{(x-1)(e^{x-1})}$ 9. $\frac{1}{xe^{5x}}$ 10. $\frac{e^x+e^{-x}}{(e^x-e^{-x})(1+\ln x)}$

Exercise 23

1. $\dfrac{dy}{dx} = 2x(2x^2 - 3x - 7)$, $\dfrac{d^2y}{dx^2} = 2(6x^2 - 6x - 7)$, $\dfrac{d^3y}{dx^3} = 2(12x - 6)$ 2. $-\dfrac{3}{x^2}$

3. $20x^2 e^{5x^4}(20x^4 + 3)$ 4. $-50x^8\sin^2 x^5 + 40x^3\cos x^5\sin x^5 + 50x^8\cos x^5$ 5. $\dfrac{128}{(4x - 10)^2}$

6. $\dfrac{-(x^2 - 2)\sin x - 2x\cos x}{x^3}$ 7. $\dfrac{4\ln x - 6}{x^3}$ 8. $-2\csc^2 x(\csc^2 x + 2\cot^2 x)$

9. $2\cos^2 x - 3\sin^2 x - 5\cos 2x$ 10. 0

11. $-8x^2\sin x^2 + 4\cos x^2 + 2\sin^2 x - 2\cos^2 x - 4x\cos x^2 + 2\cos x\sin x$

12. $\dfrac{2[(x^2\sec^2 x + 1)\tan x - x\sec^2 x]}{3x^3}$ 13. $e^{-x}(4\sin 2x - 3\cos 2x)$ 14. 0

15. $\dfrac{-(9x^2 - 2)\sin 3x - 6x\cos 3x}{3x^3}$

Exercise 24

1. $2x(3x^4 - 5)^3 + 36x(3x^4 - 5)^3$ or $2x(3x^4 - 5)^2(21x^4 - 5)$ 2. 209

3(a) $\dfrac{\sqrt{3x^2 - 1}\,(3x^2 + 2)}{x^3}$ (b) $5\sqrt{2}$ 4(a) $\dfrac{12x^2 - 3y^3}{9xy^2 - 2y - 5}$ (b) $-\dfrac{9}{16}$ 5. $-\dfrac{2}{y}$

6. $3x(x^2 + 5)^2(3x^3 + 5x - 2)$ 7. $e^{\sin x + \tan x}(\sec^2 x + \cos x)$ 8. $\dfrac{2x(\sin x - x\cos x)}{\sin^3 x}$

9. $\dfrac{2\cos x\sin x - 3\sin 3x}{\sin^2 x + \cos 3x}$ 10. $-5a^{\cos 5x}\sin 5x\ln a$ 11. $5x^2 e^{-3x} - 10xe^{-3x} + 2e^{-3x}$

12. $-(2x^3 + 2)\cos\left(\dfrac{1 - 2x^3}{x^2}\right)$ 13. $\dfrac{4x^4 - 6x^3 + 12x^2 - 18x + 6}{x^3}$ 14. 0

15. $5\left(\dfrac{2}{x^2} - \dfrac{2}{x^3}\right)\left(\dfrac{1}{x^2} - \dfrac{2}{x}\right)^4$ or $\dfrac{10(x-1)(1-2x)^4}{x^{11}}$ 16(a) $\dfrac{3x^2 e^{x^3 - y^3} - y}{3y^2 e^{x^3 - y^3} + x}$ (b) 1

17. 0 18. $\dfrac{2\sin\frac{1}{x^2}}{x^2} - \dfrac{4\cos\frac{1}{x^2}}{x^4} + 2\cos\dfrac{1}{x^2}$ 19(a) $\dfrac{2e^x(e^x + 1)}{(e^x - 1)^3}$

(b) $\dfrac{4(e^{2x} - e^{-x}) + (1 + e^{-x})(e^x - 1)^3}{(e^x - 1)^3(1 - e^{-x})}$ 20. $2\tan x$ 21. 0

22. $3\left(\dfrac{\sqrt{5}}{4\sqrt{x}} + 3\right)\left(3x + \dfrac{\sqrt{5x}}{2}\right)^2$ 23. $\dfrac{-10x - 18}{x^4}$ or $\dfrac{-2(5x + 9)}{x^4}$

24. $9x^2(x^3 + 1)^2 + \dfrac{5 + 4x - 10x^2}{(x^2 + 1)^2}$ 25. 0

Please if you found this book well simplified enough for easier understanding, kindly give it a five star rating on amazon so as to encourage people to buy this book, thereby helping them improve on their skills on coordinate geometry and differential calculus. Thank you.

If you have any enquiry, suggestion or information concerning this book, please contact the author through the email below.

KINGSLEY AUGUSTINE

kingzohb2@yahoo.com